Collecting the
SPACE RACE

Stuart Schneider

Price Guide Included

Schiffer Publishing Ltd

77 Lower Valley Road, Atglen, PA 19310

ATOMIC GUN, ca. 1957. A lithographed sparking ray gun made in Japan by HAJI, length 9 inches.

Published by Schiffer Publishing, Ltd.
77 Lower Valley Road
Atglen, PA 19310
Please write for a free catalog.
This book may be purchased from the publisher.
Please include $2.95 postage.
Try your bookstore first.

We are interested in hearing from authors
with book ideas on related subjects.

Printed in the United States of America.
ISBN: 0-88740-535-5

Title page photo:

APOLLO—X MOON CHALLENGER, ca. 1968. Battery operated, lithographed tin and plastic. The tail and nose cone light and the rocket moves on the floor, it stops, the rocket raises up and the capsule moves away from the body. It then reverses and moves along again. This example is missing the four plastic panels that open when the capsule moves out. Made in Japan by T. Nomura, 16″ long.

Contents

SPACE PORT, ca. 1953. The Planetary Cruiser Patrol Space Port could be the centerpiece of your space adventure. A group of armed spacemen protect the port against invaders. It is 10″ long and was made by Pyro in the U.S.A.

Acknowledgements

The author wishes to thank the following people who graciously agreed to let him photograph parts of their collections or contributed photographs or information, and without whose help this book would not have been possible. Jim Abicht, Atkins Enterprises, Beth Avary, Ken Aubrey, Mark Bergin, Dana Berry, John Bolton/Christmas Morning, Wayne Browning, Jack and Pam Coghlan, George Fischler, Robert Fletcher, James Hagler of the U.S. Space & Rocket Center, Kennedy Ho, Israel Levarek, Carl Lobel, Barbara Moran, George Newcomb, Lloyd Ralston, Dr. Reuben A. Ramkissoon, Jim Schleyer, Donald Sheldon, Victor Shiloski, Glen Swanson, and Catherine Saunders-Watson. With a note of special thanks to John Kleeman & Peter Kleeman whose hospitality and depth of collection added immensely to the quality of this book.

PLANET ROBOT, ca. 1957. A plastic and metal windup, sparking action robot that was probably an unauthorized "Robbie" robot. Robbie starred in the 1956 and 1957 movies "Forbidden Planet" and "The Invisible Boy". Made by K.O. (Yoshiya) in Japan, 12″ tall. Robots gave us the feeling that if we could not be there in person, we could control a mechanical explorer in our place.

Collecting the Space Race

The desire to travel into space is as old as mankind itself. There has always been a fascination with the moon and the stars and a desire to travel to distant planets. The word "planet" is from the Greek word meaning "wanderer." The roar of the rocket touches the heart. Mankind's desire to explore new places never ends. It is Man's destiny to explore and populate the stars.

It was not until the beginning of the 20th century that the idea of man traveling in space was anything more than fiction. Although hot air balloons, large enough to carry a person, lifted off the ground in the late 1700s, it was the invention of powered flight-the airplane—that gave man hope of controlling his ascent into the skies and beyond. Once man was headed up, he could never again be satisfied being tied down to the earth's surface. This desire to get into space changed the way we were used to living. The industrial revolution of the 1850s and 1860s gave man freedom by letting machines do more of his work. The space age gave man hopes and dreams of a future in the exploration of the universe. These dreams were translated into souvenirs, toys, games, entertainment and every day items with the space theme. They are the social artifacts of the space age.

What items qualify as artifacts of man's attempt to get into space? What period of time should the items cover? This volume begins with the start of the 20th century, but that is purely arbitrary. It could have begun earlier or later, but the 20th century seemed so appealing with the variety of things that inspired, compelled, and finally allowed man to explore the moon and stars.

The coverage of this volume is also arbitrary, but some restrictions had to be established and the coverage of the many fields limited. You may wonder why *Lost in Space* (1965), *Star Trek* (1966), *2001* (1968), *Space 1999* (1975), *Star Wars* (1977), *Battlestar Galactica* (1978), *The Jetsons*, *Dr. Who* and other popular space theme television shows and movies barely get mentioned. In terms of inspiring man to go further in their exploration of space, *Star Wars* and *Star Trek* have no equal. They gave us access to the images that previous generations could only dream or read about. Most, however, are fantasy adventures that are already covered in books and magazines. Toy makers have mass produced millions of toys based upon their characters and stories. Here, the intent is to concentrate on the fantasy that led up to the realities of space travel and the almost folk art aspect of the toys, games and artifacts that rode the wave of fascination with space travel. "Folk art" is used because many of the things in this volume were produced in relatively small quantities, for short periods of time, and yet, are still able to convey a sense of intimacy with the designer or artist.

Collectors of space memorabilia collect a variety of items, but many specialize and collect only a certain category. These categories include first man on the moon items, the original Mercury astronaut items, Sputniks and satellites, fantasy items (Buck Rogers, Capt. Video, Flash Gordon, Star Trek, etc.), UFOs, Ray Guns, books on space—scientific, historical or fantasy, space theme postage stamps, mission patches, autographs, and space toys and robots. Hopefully this book will inspire and answer some of the questions about the items that have been saved or collected and expose a broad field of things to collect.

SPACE PATROL FLASHLIGHT, ca. 1953. This is one of the nicest space flashlights made. It is a Ray-O-Vac product. Rather uncomfortable to hold, it looked wonderful and the box that it came in made it something that any young spaceman would want to own.

ROCKET LAUNCHING BANK, ca. 1963. Lithographed tin mechanical bank, 10 inches long, with a rocket pointed towards the moon. A coin is loaded on the back of the spaceship and fired into an opening in the moon. This is a very beautiful and rare bank showing buildings, rocket launching facilities and a rocket launching. Made in Japan.

The Dream of Space

One of the most incredible early depictions of travelling to the moon, after Jules Verne's great books, was a silent movie made in 1908 by a French movie maker, George Melies. It was called "A Trip Into The Moon". In it, six astronomers dressed in ties, jackets and top hats enter a giant bullet-shape space capsule, that is then loaded into an immense cannon. They are shot to the moon.

On arrival, they leave the space capsule carrying only umbrellas and soon meet hostile moonmen inside the moon. A short battle ensues and the moonmen are dispatched in puffs of smoke as they are struck with the umbrellas. Our heros retreat to the spaceship. Getting back to earth poses no problem. They simply climb into the capsule and one hardy soul, using a rope, pulls the ship off a cliff's edge and they fall back to earth, landing in the ocean.

One is immediately struck by the absurdity of the trip into space without spacesuits and other necessary gear. But after further review, it is amazing that this early movie maker came up with the ideas that he did, such as beings living under the surface of the moon, using a capsule-like space vehicle, and returning from the moon by landing in the ocean. At the turn of the century, few knew what to expect in space or on the moon. No one had been there. The average person thought that the moon would contain an atmosphere just like that on earth.

A noted "space expert" of this time was astronomy professor Percival Lowell. He wrote about the signals he was receiving from Mars. The canals of Mars were visible with a telescope and he stated that the Martians lived just like we did on earth. They built cities and a series of canals for transport and were signaling us that they were there. The interest in Mars was at an all time high. Man was anxious to get there and meet them. That is, until H.G. Wells' story, "War of the Worlds", written in 1898, became popular and Martians were depicted as hostile aliens.

With the coming of Halley's Comet in 1910 and the news that the earth would pass through the tail of the comet on May 18 and 19 of that year, spaceflight took on new meaning. In 1908, Professor Morehouse had discovered that the tails of comets contained cyanogen gas, the main component of the poison, Prussic acid, an active component of Cyanide. Rumors began circulating that the gas would poison the earth's atmosphere and everyone would die. Space travel now had a new purpose—to escape the impending disaster on earth. Postcards and illustrations mocked the methods of escaping the earth to find safety in space or on the moon. People floated into space in hot air balloons and dirigibles, climbed ladders to the moon, flew up in airplanes (heavier than air flight was in its experimental years in the 1910s) and were shot from cannons. Man was prepared to travel into space, but he was not quite sure what to expect once he got there.

The 1920s and 1930s ushered in a period of intense interest in space. An American scientist named Robert H. Goddard was launching and testing small liquid fueled rockets (the first on March 16, 1926) and hoping to find a way to the stars. It was he who determined that a rocket would work in the vacuum of space. Until this time, the theory was that without air to push against, there would be no way to move forward in space.

THE FIRST MEN IN THE MOON BY H.G. WELLS, May, 1958. A Classics Illustrated comic book.

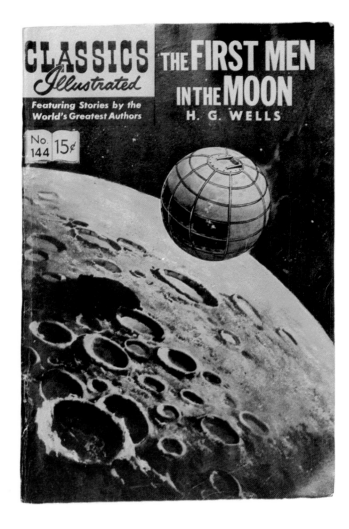

Goddard is considered to be the father of space travel, but another fictionalized character caught the attention of the public and sparked their interest in space travel. Buck Rogers, introduced in 1928 in *Amazing Stories* magazine, was the first in name recognition when it came to outer space.

Anthony "Buck" Rogers began in a Rip Van Winkle fashion. He fell into an extended sleep in a cave filled with radioactive gas and awoke in the 25th century. His 20th century skills were used to help save the citizens of earth from evildoers such as Killer Kane and the Red Mongols. Buck's 25th century companions were Wilma Deering and scientist Doctor Huer. He moved into comic strips in 1929, onto the radio in 1932 and was featured in a series of movies beginning in 1939. His adventures in space captured the imagination of millions of young, would-be spacemen.

Flash Gordon, another popular spaceman, started life in 1934 as a comic strip star. He was created as direct competition to Buck Rogers and became immensely popular in 1936-1940 when his adventures were serialized in the movies. He was an athletic star who stumbled into the laboratory of a scientist, Hans Zarkov. Zarkov claimed that disasters striking earth were directed by beings on another planet. Flash and female companion, Dale Arden, joined Dr. Zarkov aboard his rocket as it launched into space towards Mongo, the source of the destructive rays. Their nemesis was Ming the Merciless. Adventures with these space heros were the most widely distributed source of information on outer space during the 1930s.

Another source, science fiction magazines, were very popular. They contained stories and articles on space travel, space monsters and space rescues. The illustrations for the covers were done in the art style of the time and have become very collectible. Many have an Art Deco flavor. The Art Deco design style began in 1925. As the style matured, it streamlined things, adopted the sleek rocket shape and made everyday objects look like they could travel along with Buck Rogers and Flash Gordon. Vacuum cleaners, fountain pens, toasters, trains, buildings, cars and radios were all streamlined.

In the late 1930s, the outbreak of war in Europe, which became the Second World War, brought back the reality of unadorned utility to everyday objects. It also introduced the world to the German rockets that were being shot into the sky just like Flash Gordon's rocket ship. These ballistic missiles were the groundwork that would give our first space scientists real hopes of sending men into space.

After the War and into the 1950s, the interest in space travel, space exploration, and visitors from outer space was rekindled. Television was the latest miracle. The idea of seeing living, talking people on television was new to the public. New, that is, to everyone except those who grew up watching Flash Gordon. Flash's nemesis, Ming the Merciless, watched his television screen to see what was happening around him. Television shows with a space exploration theme began to appear.

Popular television shows into the mid 1950s were *Captain Video and his Video Rangers* (1949-56)(movie series in 1951), *Tom Corbett "Space Cadet"* (1950-55), *Space Patrol* with Commander Buzz Corry (1950-55) and *Rocky Jones "Space Ranger"* (1954-55).

Captain Video was the first of the television spacemen. He and his Space Rangers lived in the 22nd century and flew in a rocket ship first called the "X-9" and later the "Galaxy." A typical adventure involved an alien agent spying on a government secret invention intending to steal it. The Captain

and his Space Rangers would ultimately foil the plan.

Tom Corbett "Space Cadet" was a young cadet (top of his class) at the Space Academy in the 24th century. Tom and his two fellow cadets, Manning and Astro, would often find themselves caught up in the planning stages of an attack on some planet.

Space Patrol with Commander Buzz Corry and his slightly goofy sidekick, Cadet Happy lived in the 30th century and were kept busy flying from here to there battling powerful villains.

Rocky Jones "Space Ranger" lived in the future and due to his late entry into the television space hero business, his adventures were similar to those of the Space Patrol. Overall, each show offered the young television viewer the opportunity to step into the future for a half hour several times a week. As a result of the television shows, space toys took off in the 1950s. Toy catalogs from the early 1950s show play sets and toys based upon the show characters and themes.

Many homes did not have television in the early 1950s. If a television was not available, kids went to the movies. The movie serial, Commando Cody "Sky Marshall of the Universe" (1953) and Robby the Robot ("Forbidden Planet," 1956, and "The Invisible Boy," 1957) were at the movie theaters. The best way to get to the theater was on a bicycle and kid's bikes were again being streamlined. The headlight on the bike's handlebars was now in the shape of a rocket ship and rockets adorned the accessories. The space theme was taking off.

The interest in space touched the lives of the average man on the street. Automobiles were named as if they were rockets. Savings banks began giving out rocket and flying saucer coin banks to new customers in the 1950s. One popular style of bank was a die cast metal rocket aimed the moon. A coin was placed on the back of the rocket and shot into the moon.

At this time almost all of the space products in the United States were "Made in America," but in the late 1950s, Japanese toys began to appear in the United States. The Japanese added more whimsical decoration to their pieces. Banks in the style of American toys were also produced. A fine example of this style is the "Rocket to the Moon" mechanical bank. The base is covered in scenes of buildings, rockets being readied for launch and a rocket taking off. It was the same bank shooting a coin into the moon, but now with eye-captivating color and art work. Another desirable bank is the Astronaut Daily Dime Bank. It helped the young space aficionado to save money. The idea behind the space theme banks seem to best be summed up on one bank, "Save your pennies for a trip to the Moon."

The country was fascinated by the approaching reality of travel in space. "Out of This World" was the motto of the 1950s. The television shows and movies made space travel seem within our grasp. Space toys became very popular. By the end of the 1950s, hundreds of Japanese toys were flooding into the United States. Many were satellite, robot or flying saucer oriented as these were the themes of the movies, the news and the early space program.

In a plausible chain of events, the proliferation of space theme items made more adults aware of the impending travel into space. They voted for politicians that promised that their dreams would come true if only they would elect So-and-So to public office. Old So-and-So then supported the space programs (they were popular, did not cost much and they created jobs) and it made paying taxes seem less detestable if the money would help to get us into space.

This raises the question, "Did the 'Space' toys, games, television shows and movies mold the thoughts of the public or did they merely reflect what was happening in society?" If the question is asked in the 1960s, the answer would be that they reflected what was happening. But when the question is asked in the early to mid-1950s, the answer must be that they shaped the attitude of the public. Without the public support or "pressure" as politicians might say, there would have been little reason for the government to spend huge sums of money on the space program.

The competition for the space show was the cowboy show. Gene Autry, Hopalong Cassidy, Roy Rogers, The Lone Ranger, and others rode their horses across America. An interesting aspect of the space shows of the 1950s is that they were often cowboy shows in disguise. Good guys rode from here to there, chased bad guys or Indians, rescued the innocent and "got" the girl. Space shows were built on the Western's outline. Instead of horses they had rockets and flying saucers. The space play sets with spacemen were almost identical to western fort play sets with soldiers. In clothing, space suits were worn instead of chaps and button down plaid shirts, and ray guns were used instead of six shooters. Spacemen and cowboys were non-identical twins.

FLASH GORDON SPACESHIP, ca. 1953. The Rocket Fighter was made from the same Marx molds as the earlier Buck Rogers spaceship. Lithographed tin about 12″ long.

Space Guns

One might arguably say that ray guns have nothing to do with man's race into space since astronauts do not carry them. But ray guns are a part of space exploration. In exploring the new world in the 1400s or the western United States in the 1800s, man was armed and ready to protect himself from whatever was out there. Ray guns may be out of place in real space travel, but they were reasonable in the 1930s through 1950s. Their popularity was phenomenal and almost every adventure into space featured spacemen with ray guns. Many variations exist. There were pop guns, cap guns, water pistols, dart guns, flashlight guns, plastic and metal clickers and sparking ray guns. All were instrumental in the playful behavior of children becoming interested in outer space. Would the early space adventures be as exciting if growing something in a test tube or looking through a telescope was the main reason for venturing out into space? Ray guns helped to make it an adventure.

In sparking ray guns, the simple sparking mechanism, flint against a wheel, could be found in countless body variations. The original sparkers made in the United States in the early 1950s were heavy tin stamped models with simple graphics. The Japanese soon followed using a thinner stamped tin, making up in graphics what they lacked in durability. The tin used in these Japanese models was often scrap metal. They used whatever metal was cheap and available. Sometimes you can see other product names or art work lithographed on the inside surface of the gun. The tin sparkers are gems of the metal printer's art.

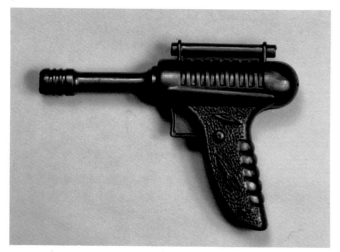

CLICKER RAY GUN, ca. 1956. Silver-Grey plastic clicker style gun, 5 inches long. Also available in blue or red. While not as exciting as the spark shooters or cap shooters, the clickers are extremely futuristic.

Die cast metal, often called "pot metal" or "white metal," was used from the late 1940s to the 1960s to replace cast iron, which had been used in making cap shooters. After the War, a new product, plastic, was used sparingly, as an accent to the metal, since it wasn't yet inexpensive. With the development of cheap plastic in the 1950s, plastic guns began to appear. The first were available as clickers, water guns, flashlights, dart guns, and some sparker models. By the 1970s, manufacturers moved production out of the United States to cheaper labor pools and began to utilize all plastic construction. Metal was discontinued due to stricter consumer products safety laws.

With water guns, the earliest models were also made of metal since plastic either did not exist or was unreliable. Examples of these early guns are the Hiller Atom Gun and the Buck Rogers "Liquid Helium" gun, which used a leather sac inside to hold the water. By the early 1950s, most water guns were plastic with a nozzle made of metal. Many of these were made in California or New York. Japanese water guns began to appear in the late-1950s and their nozzles were made of plastic. Hong Kong was the primary producer in the mid—to late-1960s while current production is almost exclusively in China. There are exceptions to these dating techniques.

These simple toys encouraged the imagination of youngsters. Of course kids shot at everything that moved, but they had to imagine that they were in space, fighting the nastiest monsters and aliens that their minds could create. While space guns were similar to cowboy guns, they were whimsical in design. A cowboy gun had to look real, while a space gun could appear in any shape. This did not stop the direct copying of styles. The toy gun collector will often find identical lithographed tin cowboy and space guns, one with a cowboy on the grips and another with a spaceship.

Space guns are collected as an art form. Most were produced for very short periods of time and were thrown out when they stopped working. The die cast metal cap shooters were usually made by companies that historically produced cast iron banks and toys. The quality of the castings are very sophisticated. One of the finest in execution is the Hubley "Atomic Disintegrator" made in the late 1940s. Hubley began making toys in the 1880s and produced the Atomic Disintegrator about 1948. They went on to produce millions of die cast cowboy style cap guns. Many space gun makers are still no more than a name or initial.

Where there are guns, there should be holsters, but space holsters are more difficult to find. There are very few to hold the numerous space guns available. The few that did survive were probably little used as they were made of thin leather or plastic with a painted design that wore quickly with even the slightest use. Buck Rogers holsters turn up more often than others.

UFOs

With all the space guns available, there had to be something out there at which to shoot. During the late 1940s through the mid-1950s, Unidentified Flying Objects (UFOs) were sighted on a regular basis. UFOs also provided another reason to explore space. There were articles in the popular magazines such as the *Saturday Evening Post* (April 30, 1949 and May 7, 1949), *True* (January and March, 1950), and *Life* (April 7, 1952). Flying saucer sightings were reported in the newspapers around the country and numerous books were published explaining and discussing UFOs. In 1950 one of these books made the best seller list—Frank Scully's, *Behind The Flying Saucers*. The book was reviewed in many magazines and as with a snowball rolling down a hill, UFO sightings increased as the flying saucer stories proliferated.

The military denied the existence of flying saucers. Private organizations were formed to investigate the UFOs, as there were many who believed that the military was not telling the whole truth to the public. Following the flurry of sightings, stories began to appear telling of human contact with the aliens. Those allegedly contacted by the aliens were soon being interviewed by the news media on a regular basis. The interviews were compelling. These were often people with impeccable credentials—policemen, pilots, housewives, elected officials, etc. Soon it was no longer "Are they out there?" but "When will they be arriving?" Groups, such as the Aetherius Society, The Inner Circle, the Unarius Foundation, and the Intergalactic Federation, were formed with the intention of providing landing sites and greeting the visitors from outer space when they arrived. Space conventions were held to bring the believers together to hear about and discuss the visitors to our planet. In 1954 the Interplanetary Spacecraft Convention held in California attracted over 5000 people. As recently as 1967, the UFO World Congress held its meeting in Mainz, Germany. The sightings were being made not only in the United States but worldwide. 1956 saw the height of the flying saucer sightings throughout the world. The interest in UFOs will most likely never die out.

One evening in 1956, a 6 year-old boy was in his backyard playing. He looked up to see a moving light in the sky. Soon other lighted objects joined the first and formed into a diamond pattern. They did not move and remained in one place in the sky. The boy ran into the house telling what he saw, but no one in his family took it very seriously. I never forgot the experience.

Toys based upon outer space, such as flying saucers, spaceships and ray guns, were finding an active buying market. The youngster of the 1950s actually felt that he was living in the space age. He or she could dress, shoot and act like a space adventurer. Then, on October 4, 1957, the World was boldly introduced to the beginning of the real space age.

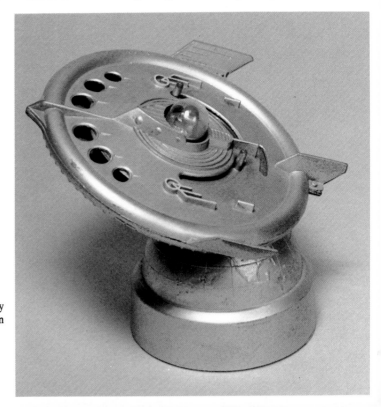

FLYING SAUCER BANK, ca. 1955. Die cast metal bank made by "Duro". Issued at the height of the flying saucer craze in America. A coin was loaded and when fired, shot around the inside of the saucer.

Sputniks and Space Exploration

"Ping...ping...ping..." "Today, October 4, 1957, the Soviet Union announced that it has successfully launched a man-made satellite, called a 'Sputnik' meaning 'Traveler', into orbit around the Earth. That sound you hear is the Sputnik's radio signal." Sputnik—the first time that we heard the word, we laughed. It seemed like a funny word when the radio announcer said it. Something to do with flying saucers or such. As the American public first laughed at hearing the word "Sputnik," the leaders of our country reacted quite differently. To them, the Sputnik was no laughing matter. To us it was the starting bell of the exploration of man's new frontier—Space.

The Sputnik announcement shocked the world. It was beyond belief that the Soviet Union could launch a 23 inch, 184 pound satellite into space and into orbit around the earth. The world was shocked because, until this announcement, the United States was the assumed leader in rocket construction. The U.S. was working on its own satellite called "Vanguard" that it had hoped to be able to launch in the latter half of 1957 in honor of the world observance of the International Geophysical Year and the public was anxiously awaiting the news of the successful launch. Delays in the program, such as test rockets which kept exploding on takeoff, pushed the expected launch date into 1958.

The Soviet Union was thought to be a country of peasants. How did they obtain the skills necessary to build a rocket and put their satellite into orbit without advance notice to the rest of the world? While they may have kept their secret from the public, it was not a secret from the military leaders of the United States. The Germans had, near the end of the second World War, designed and launched hundreds of rockets. The German scientists who built those rockets were the prizes of war that both the United States and the Soviet Union had wanted. They were captured and brought back, some to the Soviet Union and some to the United States, to work on rocket design projects. Those men were the force that launched the Space Age.

The Soviet Union's Sputnik was little more than a radio transmitter, with four antennae trailing, emitting a tracking signal. The world listened as the Russians proudly played the satellite's "ping...ping...ping..." sound at the Brussels (Belgium) World's Fair in 1958. The United States was embarrassed, furious, and its citizens were scared.

The launch of Sputnik occurred during the "Cold War" between the two super powers. Relations were poor and nuclear war seemed very close. Children of the period may remember the air raid drills that were practiced monthly in schools. "Duck and Cover" was a catchy tune that told children to hide under something and cover their eyes when the air raid siren sounded. Bomb shelters became the new neighborhood status symbols. Newspapers started saying that the Russians could launch and orbit an atomic bomb and drop it anytime and any place that they wanted. To make matters worse, there was no doubt that the Sputnik was up there. You could actually see it with the naked eye on a clear night. Any short wave radio could pick up its signal—the constant "ping...ping...ping..."

The space race had begun and souvenirs, toys and novelties were made to celebrate each new space event. Souvenirs were produced by the Soviet Union, Japan and other countries to honor the orbiting satellite. At the Brussels World's Fair in 1958, one could find novelties shaped like a sputnik—statues, music boxes that played the Russian anthem and then the satellite's "ping...ping...ping...," gold charms, puzzles, games, toys, banks, postcards, etc.

Since the first satellite did little more than emit a radio signal, its novelty quickly wore thin. Other, more interesting, things were shot into space—dogs, monkeys and then man. Souvenirs of these first Russian and American satellites were produced for a very short period of time—part of 1957 through 1960—and locating them today can be a daunting challenge. The U.S. collector of these early space items must broaden his or her search area to include the antique markets and toy shows in Europe and Great Britain.

In this race for space, a little publicized battle arose between the German scientists, led by Werner Von Braun and working for the Army, and the Department of Defense (the D.O.D.). Von Braun and his fellow scientists, known as "The Paper Clip Scientists," had been designing weapon carrying "ballistic" missiles for the Army. During their research, under the guise of launching nuclear weapons, they worked on a rocket able to launch a satellite into space. They were advised by the D.O.D. to curtail that project and stick to non-orbital ballistic missiles. The D.O.D. was so worried that the German scientists might launch a satellite into orbit that they ordered that the final stage of any rocket launched must be a dummy. In September, 1956, the Army launched a rocket that, had the final stage not been a dummy, could have easily entered orbit.

The D.O.D. under the guidance of N.A.C.A. (National Advisory Committee for Aeronautics, created in 1915, and the predecessor of NASA) was pursuing a very modest space program. The D.O.D. felt that the Navy should launch the first satellite. They believed that if the German scientists were to take this honor, the D.O.D. would lose "face" and possibly their control of space projects. They won the first round and America's first satellite, a little thing about the size of a melon, was prepared to launch on a Navy designed rocket—the

SPACE STATION, ca. 1962. Everyone is ready for the spring powered launch of the "Operation X-500" space station set. The nose cone has a dog inside and another holds an astronaut. The handsome all plastic set is 18″ long and made in the U.S.A. by Deluxe Reading. Originally came with a separate command center with missiles.

Vanguard. The rocket was launched, with full press coverage on December 6, 1957, and it exploded. Public pressure demanded that the U.S. get a satellite into space immediately. The only remaining rocket was a four stage Army Redstone rocket called a "Jupiter-C" designed by the German scientists. They were permitted to proceed and after several delays, launched America's first satellite, "Explorer 1", on January 31, 1958. Explorer 1's sensors discovered the Van Allen radiation belt about 600 miles above the Earth. If man was to go into space, the ships would need to shield him from the deadly radiation.

By the time we put Explorer 1 (6 inches wide, 6½ feet long and 30 pounds) into space, the Soviets were lifting payloads much heavier than those that the U.S. rockets could carry. They also brought added interest to space shots by launching a satellite (Sputnik II) containing a dog named "Laika" late in December, 1957. Laika stayed in space until the craft burned up reentering the atmosphere on April 14, 1958. The Soviet Union's rockets were larger. In May, 1958 they launched a payload (Sputnik III) of over 2000 pounds.

The next generation of larger U.S. rockets, which it was hoped would carry man into space, were the "Atlas" rockets. One was successfully launched in December, 1958, but the reliability of the rocket had yet to be proven. We were using our existing hardware to loft satellites into space. One of the reasons for this conservative approach was the lack of money devoted to space research. It is here that the public's fascination with games, toys and space theme entertainment helped to prod the leaders of our country to devote more money and manpower to our racing for the stars. For the public, space travel seemed real and they were ready to go. They wanted to know why the United States was dragging its feet and lagging behind in the space program.

This public pressure helped the Congress to decide to modestly increase funding for the space program. Until this time, funding for the program was minimal and was close to being cut back further. The N.A.C.A. became NASA (National Aeronautics and Space Administration) on October 1, 1958. Its directive, from President Eisenhower, was to get the United States into space as quickly as possible, still with minimal staff and money. Werner Von Braun was appointed as NASA's first director.

Eisenhower was, from an historical perspective, a conservative president. The image of John F. Kennedy, the man to follow Eisenhower, is one of an active President. Kennedy also evokes the image of the president that pushed us to land a man on the moon within 10 years. This image of Kennedy as the strong space supporter is partially an illusion. John F. Kennedy was elected President in 1960 and initially, saw no need to give NASA more money. It was expected that the U.S. would have a man in orbit some time in the coming year under the current programs and the President was concentrating on the growing problems with Castro in Cuba. Vice President Lyndon Johnson was the space supporter and was very much the man who pushed the space program. Kennedy, being the president, received the credit.

The Manned Space Race

MOON McDARE SPACEMAN, 1965. The Moon McDare Spaceman doll was made by Gilbert, the maker of erector sets. It is rumored that Moon had a few dates with Barbie but she felt that he was too far out to continue the relationship. Another doll available was a similar G.I.Joe space doll in astronaut's outfit available with a realistic space capsule. There were other astronaut action figures, the best known was Matt Mason, made by Mattel Toys. He was available with a tremendous variety of accessories. The pieces were detailed and very realistic. As with most fantasy spacemen, there was an entire arsenal of weaponry.

Before a man could go into space, NASA practically had to guarantee his safety. A man killed during the launch of a rocket would cripple the space program and generate bad publicity. Public support was necessary to pay for our advancement into space. With much secrecy, two monkeys named "Able" and "Baker" were launched aboard a rocket on May 28, 1959. Somehow the press learned that medical doctors were involved in the launch and were prepared to tell the public that we were putting a man into space. To avoid the release of this wrong information, the press was advised that the launch participants were monkeys. Unfortunately, launching the monkeys had an unanticipated, side effect on the future of the space program.

America had chosen 7 highly trained, jet test pilots to become "Astronauts"—the first men who would fly the rockets into space. They were the Mercury 7 astronauts. Cartoons and jokes began to appear picturing the American astronauts as little more than glorified monkeys (listen to the "Spaceman" track on the first "2000 Year Old Man" album by Mel Brooks and Carl Reiner or play an old Jose Jimanez comedy album). Apparently the first capsules were designed without windows and without controls. All rocket and space capsule control was originally to be made from the ground. It mattered little to the rocket scientist whether the rider was a man or a monkey. He was only along for the ride. The astronauts changed that plan. They did not intend to "ride the rocket." They expected to fly the rocket and demanded control over the space capsule's handling and design. This included putting windows in the capsule and giving override controls to the astronauts. While this seems like a small concession, the astronauts quickly learned that they were a powerful voice and could influence the space program.

On January 31, 1961, a chimpanzee named "Ham" was launched atop the Redstone rocket. This was to be the final test flight before the first U.S. astronaut went up. The Redstone rocket was a standard ballistic missile but it was reliable. NASA did not want any surprises. As we prepared to put a man into space, the Soviet Union again stunned the world on April 12, 1961 with the announcement of a successful launch of the first man in space, Yuri Gagarin, a "cosmonaut" (meaning Space Sailor) aboard the new Vostok I spacecraft. Looking back, his flight was rather incredible. The launch and orbit were fine and well thought out. The Soviets however, had no great ideas about how to bring him down. Creatively, they put in an ejection seat and on Gagarin's return, four miles above the Earth, he was ejected from the capsule and floated down separately. The capsule with its own parachute landed safely and the ejection seat idea was dropped for later flights.

Under pressure from Congress and the public, President Kennedy reluctantly agreed to give the space projects increased funding. Vice President Lyndon B. Johnson was appointed to oversee the space program. Johnson had been a strong supporter of past space program funding legislation. He was also instrumental in NASA being relocated from Florida to Texas.

With the promise of increased funding, the U.S. launched its first astronaut, Alan B. Shepard, Jr., aboard his spacecraft, "Freedom 7," on May 5, 1961. His flight lasted only 15 minutes and was reported upon by every press organization in the free world. He received a warm welcome home and was met in person by President Kennedy. The United States finally had a man "up in space" and the public interest in space flight was amplified. Virgil "Gus" Grissom was the next American in space on July 21, 1961 with "Liberty Bell 7."

Grissom, who later died in a capsule fire during the Apollo program, almost lost his life on this flight. After landing in the ocean, the capsule door blew out and water rushed in. Grissom swam out and the weight of his space suit began to pull him under. The rescue helicopter pilot mistook his waving for help to mean that he was okay and they should try to save the capsule. The capsule was lost and Grissom was saved. Rumors spread that Grissom panicked and caused the loss of the capsule. Shepard and Grissom's flights never orbited the earth and they personally never quite caught the public's admiration.

The Soviet Union, pushing ahead with its space program, launched and orbited Cosmonaut Gherman Titov on August 6, 1961. His flight lasted over 25 hours. In the field of record keeping, the Soviets were well ahead of us in every aspect of space flight—longer flights, larger rockets and larger payloads.

The astronaut that caught the American public's admiration was John Glenn. Glenn became our first orbiting astronaut in his spacecraft "Friendship 7" on February 20, 1962. He was an articulate speaker and his statements beamed with pride in America. Glenn rode atop the giant Atlas rocket and his flight was everything that the public and press wanted. The flight was delayed several times (the suspense was building) and when he finally went up he was in space longer than any other American.

An unanticipated dramatic event enhanced his flight. While he was orbiting the Earth, the sensors on the ground indicated that his heat shield had slipped out of place. The heat shield protected the capsule and the astronaut from the terrific heat generated when reentering the atmosphere. If this was true, he would be burned up on his way back to earth. The public was not told of this problem until after Glenn had safely landed. The sensors had proved wrong and the heat shield held in place.

The people of the United States were elated by this successful flight and support for the space program was at an all time high. One only had to look at the toys and games that they bought for their children or the Welcome Back Astronaut buttons that they wore. Astronaut pennants, buttons, dolls, records and souvenirs of each flight were available. Children were trying to capture the true feel of space by playing with toys that simulated rocket launches, space capsules, and other real space program events.

With the pride in America at an all time high, politicians knew that riding the coattails of this public support was good for votes. The only problem was that the Soviets seemed very much ahead of us. Our spirit of competition demanded that we come in first. After much discussion with his advisors, the President of the United States agreed that the only way that we could beat the Soviet Union in the space race was to go for the moon. The decision was made. President Kennedy announced in May of 1961 that we would put a man on the moon by the end of the decade. With this announcement came the funding for the space program. Billions of dollars (back in the days when a billion dollars was a lot of money) would have to be committed. America was now truly in the space race.

As an aside, this event was the critical factor that made the desktop computer a reality. Until this time, computers were room sized machines that were huge energy consumers and had to be insulated from shock. The transistor and integrated circuit had just been invented but their manufacture was very costly. If a rocket were to go to the moon, it would need a small, reliable onboard computer. Now that cost was no object, integrated circuits were produced and improved. The small computer was a reality and the companies that produced the integrated circuits found that they could make these chips very cheaply once the initial production machinery was built. The first pocket calculator, made by Texas Instruments with their new integrated "computer" chip, was made available to the public in 1972. More consumer products followed and the desktop computer was born.

SPACE TRAVEL COSMIC TOY WATCH, 1955. A fine little Japanese made watch for telling time in space. The graphics are excellent. With the inexpensive quartz movements in timepieces, little kids watches are now real working models.

The Race to the Moon

The souvenirs and toys produced during the space race are reminders of the joy of exploring the new frontier. "Welcome Back Astronaut" buttons and pennants were produced for many of the Mercury program space shots. The wearing of a button on your shirt or jacket showed that you cared about the men risking their lives to further our exploration of space. The Mercury 7 astronauts each took their turn on a mission (with the exception of Deke Slayton, due to a possible heart murmur) and were welcomed home as our new heros. As the Mercury program drew to a close, we prepared for the Gemini phase—two astronauts to a capsule.

Gemini was the logical continuation to Mercury. It received formal approval from NASA in December, 1961. The first two man mission was in March of 1965. Two men would handle the spacecraft and attempt new projects such as joining two spacecraft together, simulating repairs outside the craft and trying out new scientific instruments. The new, larger capsule was designed to be controlled by a pilot. The launch rocket was the now reliable Titan II. Thirty pilots in total were trained for the program. It was during the Gemini program that the first astronauts died. Elliot See and Charles Bassett, the original crew of Gemini 9 were killed when their training jet crashed. Aside from the death of a crew, Gemini flights ran effortlessly and the program was a success. There were fewer souvenirs from this period of space exploration than during the Mercury years. The items that have survived are more difficult to locate.

Two highlights of the Gemini program were the first U.S. space walk, Gemini 4, June 3, 1965, and the linkup of two spacecraft, Gemini 8, March 16, 1966. Gemini 4 was also the first flight where the astronauts were permitted to wear the American flag patch on their suits, and Gemini 5 (August 21, 1965) was the first where the first mission patches were permitted to be worn.

Even with the successes in Gemini, it seemed that the Soviet space program was always a step ahead of us. The Soviet's first space walk was by Aleksei Leonov flying with Pavel Belyayev aboard Voskhod 2 on March 18-19, 1965. The Soviets had brought two spacecraft within 3 miles of each other back in August of 1962. They put a woman, Valentina Tereshkova, in space in June, 1963 and then put three men in one capsule (October, 1964) well before the end of the Gemini program.

Current thinking suggests that Nikita Khruschchev demanded that the Soviet Union space program produce high profile missions to make the United States look weak. Each launch had to have something new and dramatically different. Once the program had reached the limits of what it could do within its abilities, it slowed. The Soviets were not ready to send a man to the moon and bring him back. With the end of the Khruschchev era, funding was reduced and the direction was changed towards studying man's ability to live in space for extended periods of time.

The Gemini program was designed to put a man into space for longer periods of time than the Mercury program. The Apollo program that followed was designed to land on the moon and return. Three astronauts to a capsule succeeded the Gemini pair. The rocket that the Apollo astronauts rode was the powerful Saturn rocket, initially designed as an intercontinental ballistic missile.

The program started off badly. On January 27, 1967, during a practice mission on the ground in the Apollo I spacecraft, a fire started in the capsule and the crew of three—Gus Grissom, Roger Chaffee, and Ed White—were dead within seconds. The fire burned quickly in the oxygen rich atmosphere inside the capsule. The launch schedule was stopped so that the capsule could be redesigned to prevent future fires. The deaths of these astronauts has made items associated with them very collectible.

Apollos 2 through 6 were unmanned flights designed to test the new systems. After the fire, the public's attention, after years of accepting the launch of a rocket as an every day affair, was renewed. With the testing finished, Apollo 7 (Shirra-Eisele-Cunningham) finally flew on a smaller Saturn 1B rocket on October 11, 1968. The flight went smoothly and set the stage for the preparation for the moon landing.

The next Apollo launch was to be an almost full dress rehearsal for the flight that would land a man on the moon. The public was beginning to get excited about rocket launches again and news coverage increased. This excitement supported a new wave of space items for the collector today. The astronauts would ride a Saturn V rocket into space and actually orbit the moon. Apollo 8 was launched on December 21, 1968 with a crew of Borman, Lovell and Anders. Americans watched as Apollo 8 went around the moon for the first time. During their 8 orbits of the moon, they were the first to witness the dark side of the moon and see an incredible "earthrise." That flight was followed by Apollo 9, the first manned flight of the lunar module, and Apollo 10, where Stafford and Cernan came within 9 miles of the lunar surface in the lunar module while pilot John Young orbited in the command module. The stage was set to put a man on the moon.

The attention of the entire world was on Apollo 11—the flight to land a man on the moon. Television covered the flight from start to finish. Everyone who could found a television set to watch the launch on July 16, 1969. People stayed glued to their television sets for the next four days. On July 20, 1969, with almost every person on earth listening to a radio or watching a television, Apollo 11's Lunar Lander with two men aboard landed on the surface of the moon. "The Eagle has landed" were the first words heard from the moon. Neil Armstrong was the first man to set foot on the moon and as he stepped from his ship, he said, "That's one small step for (a) man, one giant leap for mankind." Buzz Aldrin then joined Armstrong on the surface, Armstrong took some photographs of Aldrin and the two spent the next two hours playing and working on the moon. The event was incredible and the technology allowed the world to watch it live on television. The astronauts did not carry umbrellas and did not find moonmen running about. They confirmed that the moon was not made of cheese. This flight generated the items that most space collectors desire. What collector wouldn't walk a mile on sharp moon rocks to own a speck of actual moon dust, an autographed photo of the three astronauts, a photograph of Buzz Aldrin by Neil Armstrong, or an actual piece of their space gear. These are more than collector items, they are relics of the culmination of thousands of years of wondering about the moon.

During the moon landing, Michael Collins remained aboard the Apollo capsule orbiting the moon. The lunar lander left the surface, rendezvoused with the command module and then headed back to earth. Apollo 11 landed in the ocean and the astronauts were picked up by the U.S.S. Hornet. The world welcomed the astronauts back as heros on July 24, 1969 and promptly put them into quarantine to be sure that no moon microbes were picked up that would infect the planet. The Apollo 11 mission to the moon was completed. Man's dream to reach the moon had been accomplished. Now man could concentrate on touching the planets and someday, the stars.

Hundreds of manufacturers produced souvenirs after the moon landing. The quality of the items produced varies from incredibly cheap to enormously expensive. Phonograph records were popular. You could relive the liftoff, the touchdown, the rendezvous, and the return to earth over and over again. The records were great souvenirs. Imagine wondering what Abraham Lincoln sounded like when he recited his Gettysburg Address. We will never know, but everyone in the future will be able to hear the actual voices of the astronauts as they landed and then set foot on the moon for the first time in history, provided that someone transfers these recordings to the current music media. First day covers are available from around the world and many good books document the Apollo 11 flight.

"First Man On The Moon" commemorative items, were produced in quantity. The variety can be overwhelming to a collector so it pays to be selective. Items such as medallions, plates, drinking mugs, etc. were made by the tens of thousands and often, the item was promised to increase in value. Most of these collector pieces have not increased in value. The reason is that collector items were usually saved and kept in perfect condition. Items meant to be used were rarely saved nor are they likely to be in mint condition. Twenty, fifty or 100 years later, the collectors items will still be available in mint condition. The toys, games, and non-collector pieces which were produced in lesser quantities were thrown away and will be difficult to find. Items inspired by the flights before and after the first moon landing are usually scarcer than the moon landing pieces.

Numerous toys and games were produced that copied some aspect of the launch, the separation of lunar lander from the capsule, the landing and the return to earth. One of the more interesting items produced during this period were sculptures and paintings of the liftoff, the moon landing, the space vehicles and the astronauts. Many were painted by amateur painters who spontaneously wanted to save their impression of the event. The images communicate the emotions of the artist and his or her pride in this miracle-like event. Professional artists were on hand to record the event and many of their paintings hang in the Smithsonian National Air and Space Museum as well as planetariums throughout the country.

There is a small group of collectors who spend enormous sums for "flown items". A flown item is one that actually went into space or to the moon. Flown first day covers that have been to the moon and signed by the astronauts involved are among the most expensive. Pieces of the space capsule or space suit and even empty food cans have been saved by the astronauts or the crews that worked on the space projects. From time to time, these items appear on the market. In the past few years, one dealer offered a NASA space suit for sale at $55,000. It would make quite a centerpiece for a space collection. The most desirable flown items are from the first John Glenn flight and the Apollo 11 flight. Any item that has actually been to the moon is in hot demand.

FIRST MAN ON THE MOON COMMEMORATIVE WATCH, 1969. A small working watch made by Sheffield. The second hand is a space capsule that travels around the face.

After the Moon Landing

MOON PATROL, ca. 1958. A blue & silver lithographed metal vehicle.
Imagine driving across the face of the moon with your lighted spinning
globe, flashing windshield lights and bump and go action. Made in Japan
by T. Nomura.

The race to the moon had ended. The Soviet Union abandoned its efforts to land a man on the moon and concentrated instead upon long term living aboard a space station. Apollo continued with 6 more flights to the moon (five more landings for a total of 12 men who walked on the moon) between 1969 and 1972, where we honed our space skills and completed numerous experiments that furthered our knowledge of science and space. Man now looked to conquer the stars. Unfortunately, the Nixon administration canceled any further moon trips and the planned space station.

Programs after Apollo never generated as much excitement as the moon landing program, but they did add tremendously to our understanding of space. There also were items invented for the space program that would make our lives more comfortable in the future. The space program produced such hits as the components for the personal computer, advanced medical testing equipment, and the miniaturization of electronic

devices that appear in our cars, refrigerators, radios, and telephones.

Some highlights of the United State's progress into space after the moon landings were the unmanned Mariner 9 (to Mars, 1971), Mariner 10 (to Venus and Mercury, 1972), the Skylab programs in 1973-1974 (Life aboard a space station. Skylab I astronauts were Charles Conrad Jr., Joseph Kerwin, and Paul Weitz. Skylab II astronauts were Alan Bean, Owen Garriott, and Jack Lousma. Skylab III astronauts were Gerald Carr, Edward Gibson, and William Pogue.), Pioneer (to Jupiter, 1973 & 1974, and Saturn, 1979), the Apollo-Soyuz linkup in July, 1975 (Russian and U.S. spacecraft linked up. U.S. astronauts were Thomas Stafford, Deke Slayton, and Vance Brand. Soviet cosmonauts were Aleksey Leonov, and Valeriy Kubasov), Viking (to Mars, 1976), Voyager (to Jupiter, 1979, Saturn, 1980, and beyond) and, of course, the Space Shuttle.

The Space Shuttle

NASA went to work on the Space Shuttle when the Nixon administration cut its open ended funding. "Pick one project and stick to it," they were told. With the planned space station and further moon exploration canceled, they could send satellites to the other planets or work on what was to become the Space Shuttle. NASA chose to work on the Shuttle. The concept of the Shuttle was to be the space vehicle of the future. The first Space Shuttle launched into space was STS-1 (Space Transportation System) which flew on April 12, 1981 with John Young and Robert Crippen.

The public was again stimulated by a rocket launch. Television coverage was excellent. The rockets roar and fiery trail were inspirational. The first flight of the Space Shuttle was a success and souvenirs of our new space ship were produced. The next shuttle flight was again a test flight. Joe Engle and Richard Truly in the STS-2 lifted off in November, 1981 with further exemplary performance. Once it received the okay, it was ready for working flights.

The first satellites were placed in orbit by the Columbia shuttle in 1982. NASA felt that it was then time to begin introducing the rest of the shuttle fleet. In April, 1983 the shuttle, Challenger was launched. The shuttle Discovery followed in August, 1984 and shuttle Atlantis placed a secret military satellite into orbit in October, 1985.

The shuttle was going to be the United States' moving truck into space. Large loads could be put into orbit and then the shuttle would glide back down to earth, ready to launch again. Reality has a way of mocking the dreams of men. The shuttle was not quite so simple, was very costly and when a problem was discovered, everything shut down. The Space Agency had put all its eggs into one basket and that basket was doing pretty well until the Challenger disaster on January 28, 1986. The shuttle's main tank exploded during takeoff and the crew was killed. Prior to this, during the period between 1981 and 1986, 22 shuttle launchings occurred.

As journeys into space became more commonplace, the public seemed to lose interest in the advances of the space program. Television and press coverage was way down. It took the Challenger disaster to hammer home the idea that launching a rocket into space was an adventure, both dangerous and risky.

Without a great deal of public support, the Congress and elected officials are hesitant to spend billions of tax dollars on advancements in the space program while projects here on earth go begging for funding. The technological "profits" from the space programs have advanced science greatly. Unfortunately, the general public is not impressed by small steps. They demand quicker returns on their investments. The future of the space program is in jeopardy. Hopefully the dollars can be found to continue the exploration of space. In looking at the 500th anniversary of Columbus's voyage to the new world, people point to our accomplishments in space. Hopefully, 500 years after man landed on the moon, people will point to colonies on other planets. It is man's destiny to explore the stars and populate new worlds. To paraphrase Pogo, "We have seen the aliens from outer space and they is us."

SPACE SHUTTLE FLIGHT PATCHES, November & April, 1981. The Space Shuttle was billed as the "Space Bus". It would carry large loads and help to build space stations while launching all of our satellites. The Space Shuttle program was so promising and so costly, that there was no money available to test alternate systems. After the Challenger disaster, it became apparent that backup systems were necessary. These were the first two missions in the space shuttle.

A Brief Outline of Space History

This chart can be used to help date the different space items that may be found.

Name	Date	Event	Country
Sputnik I	Oct. 4, 1957	1st man made Satellite	U.S.S.R.
Sputnik II	Nov. 3, 1957	2nd Satellite w/Laika the dog	U.S.S.R.
Explorer I	Jan. 31, 1958	1st U.S. Satellite	U.S.A.
Sputnik III	May 15, 1958	3rd Satellite w/heavy load	U.S.S.R.
Luna I	Jan. 2, 1959	1st Satellite to escape Earth's Gravity	U.S.S.R.
Pioneer 4	Mar. 3, 1959	1st U.S. Satellite to escape Earth's Gravity	U.S.A.
Jupiter	May 28, 1958	1st U.S. Satellite w/monkeys	U.S.A.
Mercury	Dec. 4, 1959	The chimp, Ham, flies aboard the Mercury capsule	U.S.A.
Tiros 1	Apr.1, 1960	Tiros 1, 1st meteorological satellite, launched	U.S.A.
Vostok 1	April 12, 1961	1st manned Orbital flight - Yuri Gagarin	U.S.S.R.
Mercury	May 5, 1961	1st U.S. manned flight - Alan Shepard	U.S.A.
Mercury	Jul.21, 1961	2nd U.S. manned flight - Gus Grissom	U.S.A.
Vostok 2	Aug. 6, 1961	2nd manned orbital flight - Gherman Titov	U.S.S.R.
Mercury	Feb. 20, 1962	1st U.S. manned orbital flight - John Glenn	U.S.A.
Mercury	May 24, 1962	Manned orbital flight - Scott Carpenter	U.S.A.
Vostok 3	Aug. 11, 1962	2 flights together - Nikolayev	U.S.S.R.
Vostok 4	Aug. 12, 1962	2 flights together - Popovich	U.S.S.R.
Mercury	October 3, 1962	Manned orbital flight - Wally Schirra	U.S.A.
Mercury	May 15, 1963	Manned orbital flight - Gordon Cooper	U.S.A.
Vostok 6	Jun. 16, 1963	1st woman in space - Valentina Tereshkova	U.S.S.R.
Ranger 7	Jul. 28, 1964	1st rocket to moon	U.S.A.
Voskhod I	Oct. 12, 1964	Three men in one capsule	U.S.S.R.
Mariner 4	Nov. 28, 1964	1st close range phots of Mars on 7/14/65	U.S.A.
Voskhod II	Mar. 18, 1965	1st space walk - Alexei Leonov	U.S.S.R.
Gemini 3	Mar 23, 1965	1st U.S. 2 man mission - Grissom-Young	U.S.A.
Gemini 4	Jun. 3, 1965	1st U.S. space walk - White-McDivitt	U.S.A.
Gemini 5	Aug. 21, 1965	Cooper-Conrad	U.S.A.
Gemini 6	Dec. 15, 1965	Schirra-Stafford - meet w/ Gemini 7	U.S.A.
Gemini 7	Dec. 4, 1965	Borman-Lovell - meet w/ Gemini 6	U.S.A.
Diamant A	Dec. 6, 1965	France launches first satellite	FRANCE
Luna 9	Jan. 31, 1966	Vehicle landed on moon & 1st photos of lunar surface	U.S.S.R.
Gemini 8	Mar. 16, 1966	Armstrong-Scott - 1st docking in Space	U.S.A.
Surveyor 1	May 30, 1966	1st U.S. photos of lunar surface	U.S.A.
Gemini 9	Jun. 3, 1966	Stafford-Cernan - meet w/Agena	U.S.A.
Gemini 10	Sep. 12, 1966	Conrad-Gordon - Link w/Agena	U.S.A.
Gemini 11	Jul. 18, 1966	Lovell-Aldrin - Final Gemini mission	U.S.A.
Apollo 1	Jan. 27, 1967	Fire kills Grissom-White-Chaffee	U.S.A.
Soyuz	Apr. 23, 1967	Kamorov killed while landing	U.S.S.R.
Apollo 7	Oct. 11, 1968	1st 3 man mission - Eisele-Schirra-Cunningham	U.S.A.
Apollo 8	Dec. 21, 1968	1st orbit of moon - Borman-Lovell-Anders	U.S.A.

Apollo 9	Mar. 3, 1969	McDivitt-Scott-Schweickart	U.S.A.
Apollo 10	May 18, 1969	Cernan-Stafford-Young	U.S.A.
Apollo 11	July 16-24, 1969	1st men to the moon - Aldrin-Armstrong-Collins	U.S.A.
Apollo 12	Nov. 14, 1969	2nd men to the moon - Bean-Conrad-Gordon	U.S.A.
Lambda 4S-5	Feb. 11, 1970	Japan launches first satellite	JAPAN
Apollo 13	Apr. 11, 1970	Aborted landing on the moon - Lovell-Haise-Swigert	U.S.A.
Long March 1	Apr. 24, 1970	China launches first satellite	CHINA
Apollo 14	Jan. 31, 1971	3rd men to the moon - Shepard-Roosa-Mitchell	U.S.A.
Mariner 9	May 30, 1971	Mapped the planet Mars	U.S.A.
Apollo 15	Jul. 26, 1971	4th men to the moon - Irwin-Scott-Worden	U.S.A.
Prospero	Oct. 28, 1971	England launches first Satellite	U.K.
Apollo 16	Apr. 16, 1972	5th men to the moon - Duke-Mattingly-Young	U.S.A.
Apollo 17	Dec. 7, 1972	Last men to the moon - Cernan-Evans-Schmitt	U.S.A.
Skylab	May 14, 1973	Unmanned U.S. Space station launched	U.S.A.
Skylab I	May 25, 1973	Manned mission to U.S. Space station	U.S.A.
Skylab II	Jul. 28, 1973	Manned mission to U.S. Space station	U.S.A.
Skylab III	Nov. 16, 1973	Manned mission to U.S. Space station	U.S.A.
Apollo-Soyuz	Jul. 15, 1975	U.S. - U.S.S.R. link up in space	
Viking 1	Aug. 20, 1975	First US unmanned landing on Mars on 7/20/76	U.S.A.
Enterprise	Aug. 12, 1977	Space Shuttle Enterprise released in air from 747	U.S.A.
Pioneer Venus 2	Aug. 8, 1978	Landing on Venus on 12/8/78	U.S.A.
Voyager 1 & 2	1979-1986	Fly-bys of Jupiter, Saturn & Uranus	U.S.A.
STS-1 (Columbia)	Apr. 12, 1981	1st Space Shuttle launch - Young-Crippen	U.S.A.
STS-2 (")	Nov. 12, 1981	2nd Space Shuttle launch - Engle-Truly	U.S.A.
STS-3 (")	Mar. 22, 1982	3rd Space Shuttle launch - Lousma-Fullerton	U.S.A.
STS-4 (")	Jun. 27, 1982	4th Space Shuttle launch - Mattingly-Hartsfield	U.S.A.
STS-5 (")	Nov. 11, 1982	5th Space Shuttle launch - 4 crew members	U.S.A.
STS-6 (Challenger)	Apr. 4, 1983	6th Space Shuttle launch - new shuttle	U.S.A.
STS-7 (")	June 18, 1983	7th Space Shuttle launch - 1st woman - Sally Ride	U.S.A.
STS-8 (")	Aug. 30, 1983	8th Space Shuttle launch	U.S.A.
STS-9 (Columbia)	Nov. 28, 1983	9th Space Shuttle launch - Spacelab 1	U.S.A.
STS-41-B (Challenger)	Feb. 3, 1984	Manned Manuvering Unit tests, deployed Palaba B2	U.S.A.
STS-41-C (")	Apr. 6, 1984	Satellite repair in space with MMU	U.S.A.
STS-41-D (Discovery)	Aug. 30, 1984	New Shuttle, 2nd woman - Judy Resnik	U.S.A.
STS-41-G (Challenger)	Oct. 5, 1984	EVA fueling, two women - Ride & Sullivan	U.S.A.
STS-51-A (Discovery)	Nov. 8, 1984	Caught Palaba B2 & WestarVI for repair on Earth	U.S.A.

Valuing Space Memorabilia

Valuing space memorabilia can be approached several ways. In a fashion, there are several components of space memorabilia—toys & games, stamps & related philatelic material and everything else. Toy collectors have been collecting space toys for over 10 years now and there are established auction houses and toy dealers that have set prices over the years. Prices are always volatile, that is they are always changing, but they seem to trade within a range. Stamps enjoy the same established market. The non-toy/game space items have been collected for just as long a period, but rarely sell with enough frequency to set established trading ranges.

For all space items, one can use a "Supply and Demand" valuation or what would a willing buyer and a willing seller agree to as a value? An examination of the components of value are helpful. A valuable item is usually rare, but a rare item is not always valuable. A rare piece, of which only one or two are known may not have the broad appeal of a space robot of which there are 20 known. The rarity factor is not the main determinant of value.

Take the example of a collector who is attempting to find a certain item for his collection. He values that item many times higher than another person who already has one. That collector might be willing to pay $1000 for one piece, but would he buy a second or a third at the same price? It depends upon who is buying, the availability and how badly he wants the item. Due to increased competition, collectors on the East coast and the West coast of the United States often pay more than those in the central portions of the country. The Japanese are strong buyers of flown items.

Things to Consider when Collecting Space Items
Condition: Condition is the most important criteria in the field of toys and games. A toy in Mint condition with an original box, may be worth up to 5 times more than one in Excellent condition. The original box will often double or triple the toy's value. Damage to a visible part may be a major problem. An item with a piece missing may be worth 25% of one with the piece present. Repair is often possible, but to repair the item, one needs parts or another of the same with that part. Ask yourself if the price is still a bargain when you have to locate and cannibalize another piece for parts. Stamps have an established condition/value. On non-stamp, non-toy items, condition may not impart a major price difference. A mint piece may be worth no more than 2 times that of a piece in excellent condition.
Working Condition: Pieces should be in working condition if possible. While the collector who can put a piece in working

order may not care if the item needs minor attention, most collectors would rather not do that job. During cleaning or repair the piece can be broken.
Original Parts: Not all pieces need to have original parts and finishes but they are very important to some collectors. Refinishing a toy will in most instances, not increase its value and will often decrease its value. An "honest" finish or patina, indicating that the item has seen use, is preferred. An exception to this rule is large amusement rides and pedal cars. The value of these items increase with a professional refinishing.
Availability or "Will I ever find another?": Some items are always available at antique shows or specialized dealers. Ask yourself if it is just a matter of dollars to acquire an example or is this a once in a lifetime chance to find that item? With a once in a lifetime item, condition should not be the major determining factor. Many items are ephemeral or short lived. All that is left may just be a relic or remembrance of the whole item. On those items, you may never get another chance to buy one and the person behind you may be waiting for you to put it down so that he or she can buy it.
Demand: What does it mean to you? To those growing up in the period covered by this book, some items bring back a flood of memories. To some people, seeing an item is enough. To others, possessing it is required, regardless of the price. There are collectors aplenty in the field of space collecting and the asking price reflects this enormous demand.
Buy the Best: In comparing space collectibles to other collecting fields, it is expected that prices on space items will rise. Higher prices may be a blessing in disguise. Some people have no incentive to sell a piece for only a few dollars, but if they can get "a lot of money" they will sell a piece. You will pay more but you will get a piece that you may never have another chance to own. Collections can be put together for very little money, but, as has been proven true in every field of antique collecting, the best pieces in the best condition have held or increased in value at a greater rate than the more common pieces. Buy the best that you can afford. Remember, you will rarely regret having paid too much for an item, but you will always regret the good pieces that got away.
Cost: A space collection need not cost a great deal of money. Some people collect information. A collection of information is only costly in the time acquiring it. Current space items such as photographs, autographs and technical information can be gathered at very little cost from some of the sources listed at the end of this book. Joining an organization or subscribing to a magazine can provide hundreds of leads to current space items. Using your imagination, a nice collection can be put together

within your budget. Exhibiting your collection at a local library can generate new leads to the type of material that you collect.

Investment: I discourage people from "investing" in collectibles. Collecting should be for fun and not just profit. Profits will be come if you collect the items that you enjoy and at some later time decide to sell. Buying for investment only, means never playing with your toys. They could get scratched or broken. Buying, selling and trading items can help you hone a collection to those items that bring you the most pleasure.

Locating items: Space items can be found at garage sales, antique stores, antique toy shows, flea markets, mail auctions and through ads in the antique collector newspapers. A good start may be to run an ad in your local newspaper or do an exhibit of your collection at a local library. Both of these should provide you with numerous leads. Try to meet other collectors, they can provide leads to sources that you have not considered and they provide potential trading partners.

It is often said that a price guide is out of date the moment that it is published. Do not let that affect your use of a value guide. A value guide is comparative. It allows you to compare two items to determine if they are of comparable value. It is useful in buying and trading and it can help give you a feel for the rarity of a piece.

Remember, two is a coincidence, three is a collection. Happy hunting!

SPUTNIK BEANIE, 1957. A wonderful beanie made with springs and bells so that you too could look and sound like the sputnik. The sputnik beanie is an incredibly rare item that would have been made, at best, for less than one year.

ROCKET SHIP FIRECRACKERS, ca. 1933. A rare brick (80 packs of 16 firecrackers each) label with a great Buck Rogers type rocket. Firecrackers had been imported from China since the late 1800s. In the 1920s, the manufacturers began to market firecrackers with fancy art work on the labels in an attempt to create name or event recognition. Labels were produced to honor baseball players, aviators, animals, popular movie themes, etc. Space exploration was a popular theme in the 1930s and the labels reflect the public's interest.

The Gallery of Collectibles

BANKS

STRATO BANK, ca. 1955. Die cast metal bank, 2 inches x 3.5 inches x 8 inches long, with a rocket pointed towards the moon. A coin is loaded on the back of the spaceship and fired into an opening in the moon. This was a popular bank that was given away to people opening new accounts at local savings banks. It can be found with different bank stickers on it. Usually it is colored light green. Made by Duro Mold & Mfg. Co., Detroit. There were numerous die cast metal banks made in the shape of rockets in the mid to late 1950s.

ROCKET BANK, ca. 1956. Die cast metal bank. A coin was loaded and fired into the top of the very sleek rocket.

ROCKET BANK, ca. 1954. An impressive red and yellow plastic bank in the shape of a Space Patrol rocket made by Fosta.

ROCKET BANK, ca. 1956. Die cast metal bank made by "Duro". A coin was loaded and fired into the front of the bank.

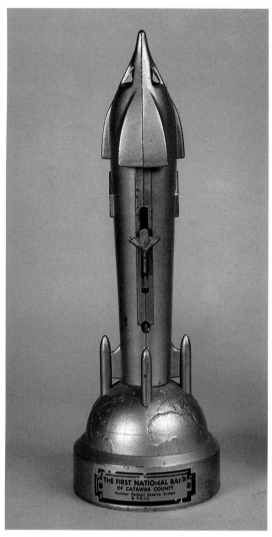

SPUTNIK BANK, ca. 1958. Lithographed tin bank in the shape of a satellite with the face of the Soviet Union's space dog, Laika, looking out.

SPACEMAN BANK, ca. 1957. Lithographed tin bank in the shape of a rocket with a spaceman shooting at flying saucers.

SATELLITE BANK, ca. 1960. Plaster bank in the shape of a satellite marked "Space Bank".

MOON BANK, ca. 1960. Plaster bank in the shape of the moon. The wording reads "Save your pennies for a trip to the moon".

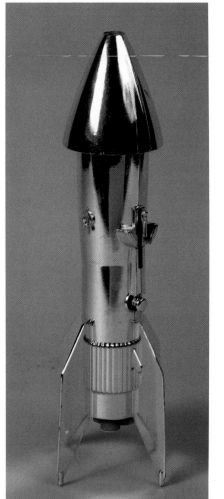

ROCKET SHIP MECHANICAL BANK, ca. 1957. Chromed die cast metal in the shape of a rocket made by the Astro Co. of E. Detroit. Marked "A Berzac Creation". A trigger and spring powered piece on the side shoot nickels and pennies into the nose cone. The same rocket can be found in a silver painted model as well. Originally it had a rubber nose cone tip and tiny rubber tips at the bottom of the fins. 3″ x 13″ tall.

SPACE CAPSULE BANKS, 1961 & 1962. Soft silver colored plastic bank issued in honor of Alan Shepard's flight. The bottom has an illustration of an astronaut on his back in the capsule and the flight path and data about Shepard's flight. The same in a bronze colored plastic with the added inscription for John Glenn's flight.

SPACE KING BANK, ca. 1960. Plaster bank in the shape of a spaceman with a mustache. The added charm of this bank is that the head is mounted on a spring which makes the Space King nod when you put in a coin.

ASTRONAUT DIME BANK, 1962. Lithographed tin metal bank 2.5 inches high. Made by Kalon Mfg. Corp.

SUN & PLANETS BANK, ca. 1960. A large metal bank showing the solar system with rubber planets.

DONALD DUCK BANK, 1966. Plastic bank in the shape of spaceman Donald Duck. Originally held Nabisco "Puppets" cereal.

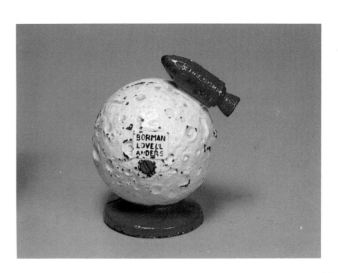

SPACE CAPSULE BANK, 1962. Die cast metal bank issued in honor of the first Mercury flights.

APOLLO 8 BANK, 1969. Cast iron bank in the shape of the moon with the Apollo 8 ship. The lettering reads "Borman, Lovell, Anders". It is unusual that a cast iron bank would be made for a specific flight, albeit an important one.

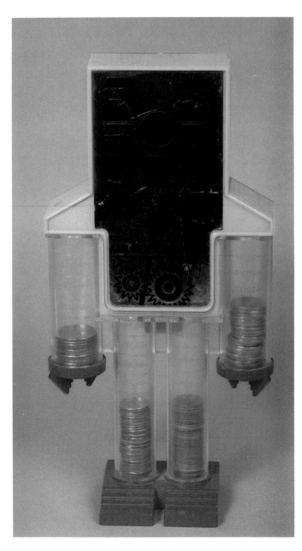

MAN ON THE MOON BANK, ca. 1969. A painted plaster bank issued shortly after the moon landing showing an astronaut planting a flag on the moon.

ASTRONAUT BANK, ca. 1972. A painted plastic bank issued after the moon landing.

MR. ROBOT BANK, 1965. Plastic bank in the shape of a robot. The Robot sorts the coins as they fall through his body.

ASTRONAUT BANK, ca. 1981. A painted plastic bank issued after the
first shuttle flight. Made in Hong Kong.

SPACE CADET PIN, ca. 1952. A celluloid pin with the Space cadet insignia. It is 2.5 inches wide.

HALLOWEEN FESTIVAL BUTTON, 1953. The Anaheim, California Halloween Festival's 1953 theme was "Out of this World" and showed a witch riding a rocket to the moon. The image conveys the public's mood at the time.

RUSSIAN ENAMEL LAPEL PINS, 1959. The gold on white pin was for the Luna 1 launch on January 2, 1959. This was the first space vehicle to achieve enough speed to break free of the earth's gravity. It was headed for the moon but due to a slight error, ended up orbiting the sun instead.

SPUTNIK PIN, ca. 1957. A nice 1″ pin to tell the world that you saw the Sputnik.

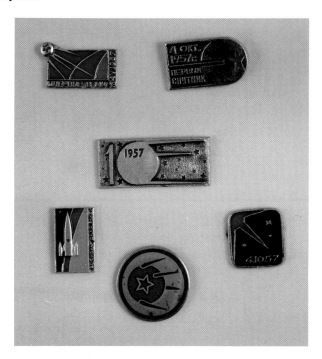

RUSSIAN ENAMEL LAPEL PINS, 1957. The lapel pins were given to those working on or connected with the different Soviet space programs. This group is for the Sputnik program. The 1950s and 1960s Soviet pins are hard to obtain. Be warned that new pins are being produced in a similar style that honor the 20th and 30th anniversaries of different Soviet space events.

RUSSIAN ENAMEL LAPEL PINS, 1959. These are rare pins for the advisors to the Soviet space program.

RUSSIAN ENAMEL LAPEL PIN, 1959. This is a rare pin for the workers at the Soviet space center at Baikonour.

RUSSIAN ENAMEL LAPEL PINS, 1959. These were for the Luna launches in 1959.

RUSSIAN ENAMEL LAPEL PINS, 1960. Three pins created for the Vostok program workers.

RUSSIAN ENAMEL LAPEL PINS, 1961. A group of pins created for the Vostok 2 flight on August 6, 1961 of Gherman Titov.

RUSSIAN ENAMEL LAPEL PINS, 1961. A group of pins honoring Yuri Gagarin's flight on April 12, 1961.

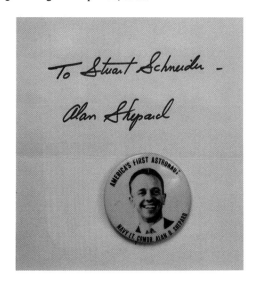

ALAN SHEPARD AUTOGRAPH & BUTTON, May 5, 1961. "America's 1st Astronaut, Navy Lt. Cmdr". There are very few items made honoring Alan Shepard's flight even though it was the first manned U.S. flight. The button can be found in two sizes.

JOHN GLENN BUTTONS, February 20, 1962. John Glenn's flight was the first American to orbit the earth. His predecessors' flights were short "shoot them up and bring them down" type flights. Glenn had captured the public's attention. At the astronaut press conferences he was very well spoken and intensely patriotic. He became more than a highly trained test flight pilot, he was the image of the American Astronaut.

RUSSIAN ENAMEL LAPEL PINS, 1962. This group honors the first fly-by of two capsules together in August, 1962—Vostok 3 and 4.

WALLY SCHIRRA, October 3, 1962. A hard to find Schirra button for his flight around the earth. 3.5″ in diameter.

SCOTT CARPENTER, May 24, 1962. The men, like Scott Carpenter, who followed John Glenn, were given hero's welcomes when they returned, but they paled in comparison to the honors bestowed upon Glenn. 1.5″ and 3.5″ in diameter.

RUSSIAN ENAMEL LAPEL PINS, 1965. Two pins for the first space walk of Alexei Leonov on March 18, 1965.

GORDON COOP-ER, May 15, 1963. Cooper's flight was the first two day flight. This button is rather hard to find. 1.75″ in diameter.

GEMINI, 1965. The Gemini flights were made with two astronauts on board a larger spacecraft. The typical flight did not generate the tremendous interest that the Mercury 7 flights received. Souvenirs created were usually for the most newsworthy flights such as the first Gemini flight (Grissom-Young, March 23, 1965). 3.5″ in diameter.

GEMINI, 1965-1966. Souvenirs of the first space walk (McDivitt-White, June 3, 1965) and the first rendezvous of two capsules in space (Borman-Lovell/Schirra-Stafford, December 4 & 15, 1965). 3.5″ in diameter. Mission patches were not worn during the Mercury program and the first three Gemini launches. With Gemini 4, NASA allowed the astronauts to wear an American flag patch on their suits. Gemini 5 (Cooper-Conrad, August 21, 1965) was the first flight where a mission patch was worn. The design on that patch was the conestoga wagon.

APOLLO 8, December, 1968. Apollo 8 buttons. One and three quarter inches in diameter.

APOLLO 8, December, 1968. Apollo 8 buttons. 3.5 inches in diameter.

APOLLO 7 & 9 FLIGHT PATCHES, 1968 & 1969. Apollo 7 was the first three man mission to orbit the earth. Apollo 1 had suffered a fire on the ground killing the crew and the next 5 Apollo missions were unmanned test flights. The Apollo 9 was the first to test the lunar lander.

MAN ON THE MOON BUTTON, ca. 1969. A 3-D style button showing the men planting the flag on the moon with the earth in the background.

APOLLO 8 & 11 FLIGHT PATCHES, 1968 & 1969. Apollo 8 was first to orbit the Moon. Apollo 11 of course, was the first manned landing on the Moon. Flight patches are fun to collect. They are readily available and have been reproduced by the millions. Most flight patches are currently available at the Smithsonian Air and Space Museum in Washington, D.C.

APOLLO 11, July, 1969. Apollo 11. 1.75″ diameter buttons with ribbons.

APOLLO 11, July, 1969. Apollo 11. 3.5 inch "Life" button and 1.75″ diameter "feet on the moon" button.

APOLLO 12, November, 1969. The second landing on the moon (Conrad-Gordon-Bean, November 14, 1969). A rare button, 1.75 inches in diameter.

APOLLO 11, July, 1969. Apollo 11 was the first manned landing on the Moon. 1.75″ diameter buttons with ribbon variations. This button, without the ribbon, has been reproduced in this size and the larger size and is currently available at the Smithsonian Air and Space Museum in Washington, D.C. for about one dollar. The originals will show their age even if in perfect condition.

APOLLO 11, July, 1969. Apollo 11 buttons. 3.5″ diameter buttons.

APOLLO 12 & PROGRAM FLIGHT PATCHES, 1969 & 1967. Apollo 12 was the second crew to land on the moon. Program patches were the general patches created for the entire program.

APOLLO 11, July, 1969. Apollo 11 buttons. 3.5″ diameter buttons.

APOLLO 16, April, 1972. The Apollo 16 is a rare button honoring Duke, Mattingly & Young for their April 16, 1972 flight to the moon.

APOLLO 10 & 14 FLIGHT PATCHES, 1969 & 1971. Apollo 10 flew around the moon, further testing the lunar lander. Apollo 14 was the third to land on the moon.

SKYLAB 1 & 2 FLIGHT PATCHES, May & July, 1973. Skylab was America's first space station. The Skylab programs did not generate great public interest, but they were our first attempts to spend extended time in space while doing scientific research.

SPACE SHUTTLE FLIGHT BUTTONS, April, 1981 & March, 1982. The second and third mission of the space shuttle. 3.5 inches in diameter.

SPACE SHUTTLE FLIGHT BUTTON, November, 1981. The first mission of the space shuttle. 3.5 inches in diameter.

RUSSIAN SPACE PINS, 1980s. The Soviet Union reissued and produced space pins to commemorate 20th, 25th & 30th anniversaries of memorable space events. These were produced for collectors living in the Soviet Union.

REX MARS PLANET PATROL, ca. 1953. 8″x 12″ Cloth flag with rockets. Perfect for a clubhouse or to attach to the back of a bicycle. Young space adventurers could identify other Planet Patrollers when their flag was flying. There were not many items made for Rex Mars.

SPACELAND, ca. 1954. Felt pennant with rocket. Lindberg started his flight to Paris from Roosevelt field in 1927. Someone set up a space theme amusement park or camp in the early 1950s at the field.

CAPE KENNEDY PENNANT, ca. 1965. Felt pennant with a multi-colored illustration of a rocket taking off. One of the most attractive pennants.

GEMINI IV PENNANT, 1965. A rare and unusual pennant for our first space walk.

SATURN V PENNANT, 1968. A Kennedy space center pennant.

APOLLO 11 COMMEMORATIVE PENNANT, 1969. Pennants were created in honor of our landing a man on the moon. This one shows the men picking up rocks on the moon with moon rock "pooper" scoopers.

APOLLO 11 COMMEMORATIVE PENNANT, 1969. A medallion style design in two color variations.

APOLLO 11 COMMEMORATIVE PENNANT, 1969. A real photo style pennant that is rather rare.

BUCK ROGERS SPACE GUN, ca. 1934. A heavy "blued" metal ray gun made in the United States by Daisy Manufacturing, length 10.5 inches. It was introduced as the model XZ-31 and was soon called the "Rocket Pistol". The grip pumps the action and the gun pops when the trigger is pulled. This is one of the classics and was available in this standard size and a smaller version called the XZ-35.

AMAZING STORIES, 1951. Spaceman, firing two ray guns, defends a beautiful spacegirl from an unseen enemy. This was the typical spaceman image that the children of the 1950s grew up with.

BUCK ROGERS SPACE GUN AD, ca. 1934. The "Rocket Pistol".

RAY GUN, ca. 1936. A stamped metal (pop gun) ray gun made by All-Metal Products, Wyandotte, Michigan, length 7 inches. It used a captive cork to make the pop. A solid compact ray gun that could be knocked to the ground in a struggle and still be relied upon to zap the aliens when you recovered it.

BUCK ROGERS DISINTEGRATOR PISTOL, ca. 1936. A heavy copper plated metal ray gun made in the United States by Daisy Manufacturing, length 10.5 inches. The Disintegrator was introduced about 1935 and was available into the mid-1940s. A spark was produced in the window on the top of the gun when the trigger was pulled. This is the classic upon which other ray guns were based. Available in several finishes—the nickel or copper models with 4 flutes on the barrel (1930s models) called the "Disintegrator" or XZ-38 and the Blue or Gold models with 3 flutes on the barrel (1940s models) called the "U-235" or "Atomic Pistol". The polished copper is the most striking. It cost fifty cents when introduced.

BUCK ROGERS LIQUID HELIUM WATER GUN, ca. 1936. A painted, stamped metal water gun made in the United States by Daisy Manufacturing, length 7.5 inches. It was also called the model XZ-44. The metal body had a leather sac to contain the water. Plastic was not made in the 1930s. The XZ-44 was later available in a bronze finish and the "Liquid Helium" name was dropped. In perfect condition, the colors and patten on this gun make it a gem in any collection.

BUCK ROGERS HOLSTER, ca. 1936. for the Disintegrator Ray Gun.

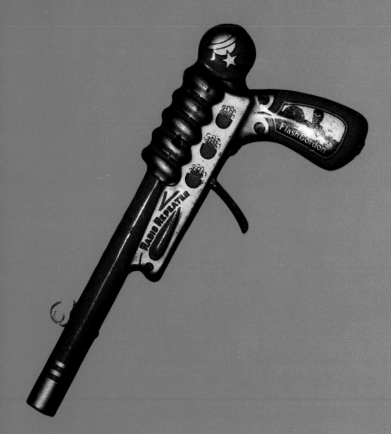

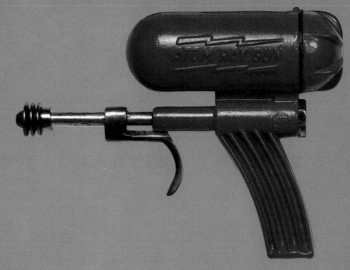

ATOM RAY GUN, ca. 1948. The Atom Ray Gun water pistol was precision made and very futuristic. Built of aluminum and brass in the United States by Hiller Mfg., it is only 5.5 inches long. The tank on top holds the water. These are usually found in poor condition with the aluminum corroded and are rare in good condition. It seems strange to think of a watergun being made in metal, but plastics were in their infancy at the time. Another metal water ray gun was the "Atom" made in the 1950s.

FLASH GORDON RADIO REPEATER RAY GUN, ca. 1937. A lithographed tin clicker ray gun, length 10.5 inches. The Radio Repeater has a 1935 copyright date, but the date probably refers to the Flash Gordon character. The same body molds were used to make the later Tom Corbett space gun.

ATOMIC DISINTEGRATOR RAY GUN, ca. 1949. A die cast metal cap shooting ray gun made in the United States by Hubley, length 8 inches. Hubley Manufacturing Company of Lancaster, Pennsylvania was started in the late 1880s and produced iron toys until 1948 when they shut down their cast iron foundry. The company continued to produce die cast white metal cap guns (mostly cowboy styles) throughout the 1950s. Hubley's high quality casting expertise is apparent in this model. Its owner must have been the envy of the neighborhood space patrol. Bright red handles, dials and thing-a-ma-bobs make this one of the most striking ray guns to be found.

BUCK ROGERS HOLSTER, ca. 1942. for the Atomic Pistol.

DAISY ZOOKA POP PISTOL, ca. 1952. A lithographed sheet metal ray gun made in the United States by Daisy using their original 1930s Buck Rogers pop gun molds, length 7 inches. The handle cocks the gun to make it pop.

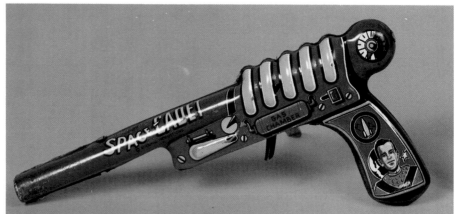

TOM CORBETT SPACE CADET SPACE GUN, ca. 1952. A lithographed sheet metal clicker ray gun made in the United States by Marx using their original 1930s Flash Gordon "Radio Repeater" gun molds, length 10.5 inches.

PLANET CLICKER BUBBLE GUN, ca. 1953. A plastic gun made by Mercury Toys of Brooklyn, New York, length 8 inches. By dipping the front of the barrel in the bubble solution, each squeeze of the trigger produced a bubble and a click. A similar model, called the "Clicka Bubble Shooter" was made by Ranger Steel Products of Long Island, New York. An earlier metal bubble shooter called the "Atom" was produced (maker unknown) in the late 1940s.

COSMIC RAY GUN, ca. 1953. A plastic sparking ray gun made in the United States by Ranger Steel Products Corp., length 8 inches. It has a replaceable flint and has a pleasing futuristic look.

SPACE NAVIGATOR GUN, ca. 1953. A lithographed tin sparking-style ray gun made in the Japan by Asahitoy. This is an early Japanese piece with great but slightly crude graphics. It is a preview of the graphics to come in the later 1950s and 1960s. Length 3.5 inches.

FLASH GORDON SIGNAL PISTOL, ca. 1953. A green painted sheet metal ray gun made in the United States by Marx, length 6.5 inches. It made a siren-like noise and shot sparks. This is one of the nicest Flash Gordon guns and rare with decals intact.

COSMIC RAY GUN, ca. 1954. A lithographed tin sparking ray gun made in the United States, length 9 inches. American made ray guns were built from heavier metal than the Japanese models and the sparking flint was usually replaceable. Colors were bold but the art work was much simpler than the Japanese guns that followed. The art work on the boxes was also much simpler. This gun was probably made by Ranger Steel Products Corp.

SPACE PATROL, 1954. Plastic ray gun that shoots rubber tipped darts.

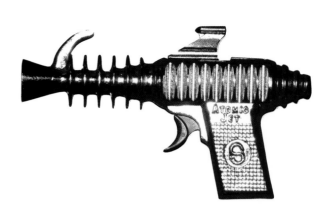

SPACE CONTROL SPACE GUN, ca. 1954. A lithographed tin sparking ray gun made in Japan by T. Nomura, length 3 inches.

ATOMIC JET GUN, ca. 1954. A die cast metal cap-shooting ray gun made in the United States by J. & E. Stevens, Cromwell, Connecticut. Stevens was started in 1843 and began by manufacturing hardware such as hammers and axe heads. This is an unusual gun in that the body was gold colored, length 8.5 inches.

ROCKET POP GUN, 1955. Pump it to pop it. Made of wood.

ATOMIC FLASH GUN, ca. 1955. A lithographed tin sparking ray gun made in the United States by J. Chein & Co., length 7.5 inches. One of the biggest maker of American tin toys in the 1940s-1950s was J. Chein & Co., started in New York City about 1900. They moved to Burlington, New Jersey in 1949.

STRATO GUN, ca. 1955. A die cast metal cap shooting ray gun in chrome, length 12 inches. Made by the Futuristic Products Co., Detroit, Mich.

888 SPACE GUN, ca. 1955. A lithographed tin cap shooting space gun made in Japan, length 3 inches.

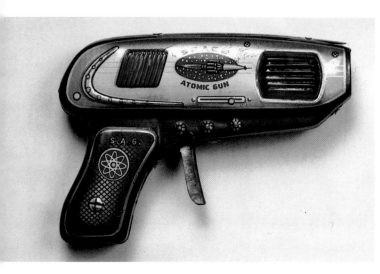

SPACE ATOMIC GUN, ca. 1955. A lithographed tin sparking ray gun made in Japan, length 5.5 inches.

SPACE GUN PUZZLE KEY CHAIN, ca. 1956. Each ray gun came apart in a secret way. Made in the United States by the Plas-Trix Co., Brooklyn, New York.

SPACE GUN, ca. 1955. A lithographed tin sparking ray gun made in Japan by "San", length 3.25 inches.

RADAR GUN, ca. 1956. Burgundy plastic clicker style gun with a green spaceman on top. 5.5 inches long.

SPACE CONTROL RAY GUN, ca. 1956. Red plastic clicker style gun. 5.5 inches long.

SUPERIOR ROCKET GUN, ca. 1956. Grey plastic dart gun. This would have gone with the lithographed tin space targets illustrated elsewhere, 8 inches long.

IDEAL FLASH GUN, ca. 1957. A plastic flashlight ray gun made in the U.S. by Ideal, length 9 inches.

X100 MYSTERY DART GUN, ca. 1956. Yellow plastic dart gun. The unusual feature is that the dart tips have tiny holes and the heads are filled with talcum powder, so there is a puff of smoke when the dart hits. Also available in grey. Made by Arliss Co., Brooklyn, New York, 3.75 inches long.

SPACE GUN, ca. 1957. A lithographed tin sparking ray gun made in Japan by Daiya, length 6 inches. When the trigger is pulled, sparks shoot out the barrel

SPACE GUN (with Rocket blasting Off), ca. 1957. A lithographed tin sparking ray gun made in Japan by S. Yoshiya, length 7 inches.

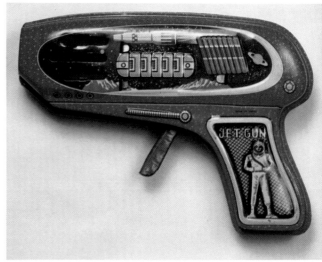

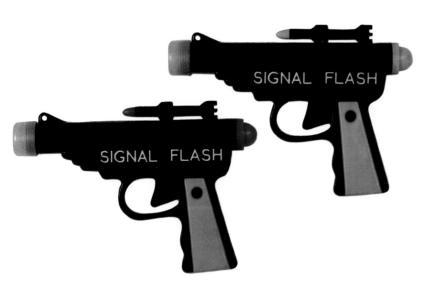

JET GUN, ca. 1957. A lithographed tin sparking ray gun made in Japan, length 6 inches.

SIGNAL FLASH GUN, ca. 1957. A plastic flashlight ray gun made in the United States, length 6 inches.

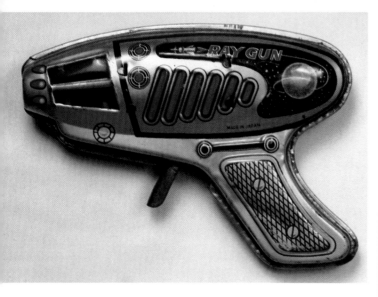

RAY GUN, ca. 1957. A lithographed tin sparking ray gun made in Japan, length 6.5 inches.

SPACE WATER GUN, ca. 1957. Red plastic water gun. Water pistols were sold by the millions every spring and by the end of the summer, they were usually broken and forgotten. This example is 5.5 inches long and made by Palmer Mfg. in the United States. The earlier space water pistols were made in the United States. During the late 1960s to 1970s, they were made in Hong Kong and current production is made in China.

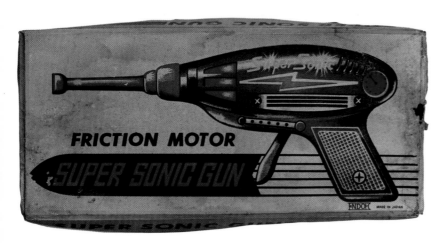

SUPER SONIC GUN, ca. 1957. A lithographed tin sparking ray gun made in Japan by Endoh, length 9 inches. Sparks shoot out the barrel when the trigger is pulled.

SPACE JET WATER PISTOL, ca. 1957. Black plastic water gun. 4 inches long and made by Knickerbocker, N. Hollywood, Calif.

SPACE JET GUN, ca. 1957. A lithographed tin sparking ray gun made in Japan, length 9 inches.

ATOMIC RAY GUN, ca. 1957. "Captain Space Solar Scout" plastic flashlight and sound ray gun rifle made in the United States by Marx, length 30 inches.

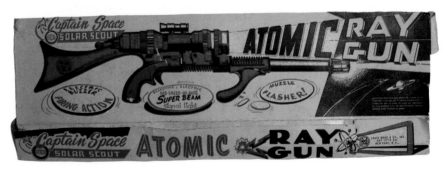

SPACE GUN, ca. 1957. A lithographed sparking ray gun made in Japan, length 9 inches.

SPACE ROCKET PISTOL BOX, ca. 1957. A tin battery powered sparking ray gun made in Japan by T. Nomura. Battery powered guns were unusual as most were primarily flashlight styles.

S-58 SPACE GUN, ca. 1957. A lithographed tin sparking ray gun made in Japan, length 12 inches.

SUPER SONIC SPACE GUN, ca. 1957. A lithographed tin sparkling ray gun made in Japan by Daiya, length 7.5 inches.

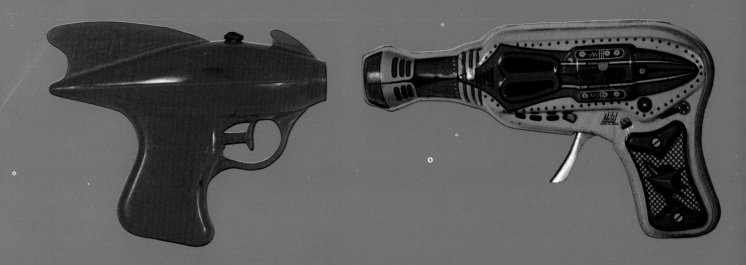

ROCKET JET WATER PISTOL, ca. 1957. Red plastic water gun. 5 inches long and made by U.S. Plastics of Pasadena, California.

SPACE GUN, ca. 1960. A lithographed tin sparking ray gun made in Japan by Hero Toy Co., length 7 inches.

JACK DAN SPACE GUN, ca. 1959. Die cast metal cap gun air brushed blue. 7.5 inches long and made in Barcelona, Spain by Metamol. The gun is also found in red. In design, it is similar to a Dan Dare cap shooter ray gun. Dan Dare was a space character in Great Britain similar to Buzz Corry or Tom Corbett in the U.S.

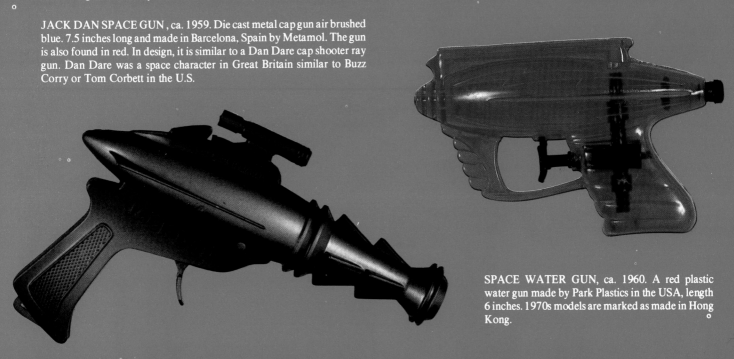

SPACE WATER GUN, ca. 1960. A red plastic water gun made by Park Plastics in the USA, length 6 inches. 1970s models are marked as made in Hong Kong.

SPACE ATOMIC GUN, ca. 1960. A lithographed tin sparking ray gun made in Japan by "T", length 4 inches.

51

SUPER SPACE GUN, ca. 1960. A lithographed tin sparking ray gun made in Japan, length 6 inches.

CLICKER RAY GUN, ca. 1960. These under appreciated plastic clicker guns were nicely made and came in wonderful shapes and styles. This example made by Irwin Toys is 9″ long and is a more unusual two-tone plastic.

SPACE OUTLAW RAY GUN, ca. 1965. The Space Outlaw was made in England and is among the nicest looking space cap guns made. It is chrome plated, die cast metal, 10 inches long. The barrel moves in and out as the trigger is pulled. Dials near the trigger turn to allow the user to use "Cosmic", "Sonic" or "Gamma" power levels. There was a matching plastic water gun made in Hong Kong during the late 1960s to 1970s.

SPACE GUN, ca. 1967. A Japanese made (by Shudo) lithographed tin sparking ray gun. 4 inches long.

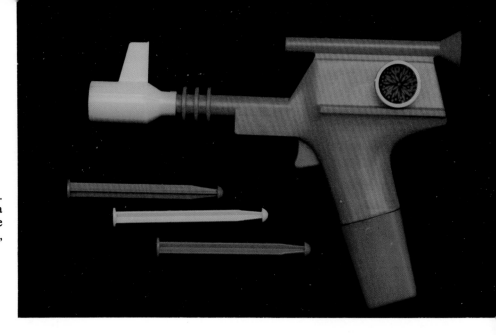

RAYDART GUN, ca. 1968. Blue plastic dart gun with a secret compartment in the handle. Made by Tarrson Co., Chicago. 9.5 inches long.

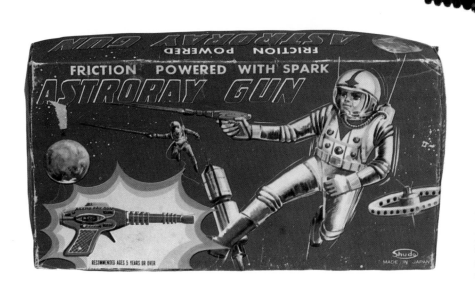

ASTRORAY GUN, ca. 1968. A Japanese made (by Shudo) lithographed tin sparking ray gun that was produced well into the 1970s (late 1970s model came in a wedge shaped box). Although a later model, this gun holds true to the form of the sparking ray guns of the late 1950s. 9 inches long.

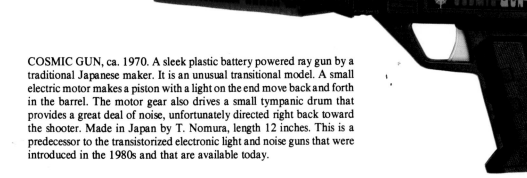

COSMIC GUN, ca. 1970. A sleek plastic battery powered ray gun by a traditional Japanese maker. It is an unusual transitional model. A small electric motor makes a piston with a light on the end move back and forth in the barrel. The motor gear also drives a small tympanic drum that provides a great deal of noise, unfortunately directed right back toward the shooter. Made in Japan by T. Nomura, length 12 inches. This is a predecessor to the transistorized electronic light and noise guns that were introduced in the 1980s and that are available today.

STRATO-BLAZER FLASHLIGHT, ca. 1933. One of the earliest "Man in Space" flashlights à la Buck Rogers and Flash Gordon. If you were going to travel in space, you would need a good flashlight to light your way on the dark side of the moon. The name "Flashlight" came from the fact that the earliest models (about 1899) could only be turned or flashed on for a few seconds. Anything longer would instantly drain the battery. By the time this model was introduced, a user could get 20 minutes of light from one set of batteries.

SPACE CADET SIGNAL SIREN-LITE, ca. 1953. 7½ inches long with a siren in the end. Made by Pifco.

CAPTAIN RAY-O-VAC-ROCKET SHIP, ca. 1952. Captain Ray-O-Vac adorns this flashlight whose box punches apart and assembles into a rocketship. It came with a full color adventure and game comic book.

TOM CORBETT SPACE CADET ROCKET LITE, ca. 1953. Push the nose cone and the red tail lights up and casts its beam. A pin on the back lets the lucky cadet wear this light on his or her shirt. Made by USAlite.

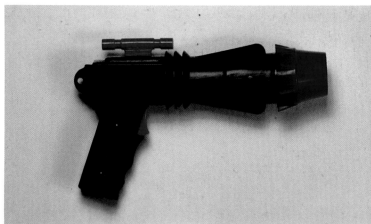

SPACE RANGER JET AUDIO-LITE, ca. 1954. Two versions—7½ inches long with a siren in the end and 7 inches long with a clicker on the side. There are planets and rocket ships on the other side.

BUCK ROGERS SONIC RAY GUN FLASHLIGHT, ca. 1955. A plastic battery powered flashlight 7½ inches long. Is it a ray gun or a flashlight? The plastic used to make this ray gun flashlight is somewhat unstable and oxidation causes a white film to appear on the surface.

DELTA BICYCLE LIGHT, ca. 1957. A bike light made of plastic and metal, 9 inches long in the shape of a rocket. In the mid to late 1950s, this was the light to have on your bike.

SPACE BOY SIREN-LITE, ca. 1954. A flashlight, 7 inches long with a siren in the end and a clicker code key for flashing the light to send signals. There are planets, mountains, craters and rocket ships on the reverse side.

SPACE SATELLITE FLASHLIGHT, ca. 1957. A satellite designed to be worn on your shirt. When the string was pulled, the light lit so that you could send signals just like the U.S. satellite that was planned to launch later that year. Made by Micro-Lite.

SATELLITE WRIST FLASH-LITE, ca. 1958. These little beauties were great fun on summer nights when we played in the neighborhood. They were worn on the wrist. When the button was pushed, the light lit. Little red and green filters slid in front of the light path so that you could signal your friends. Made by Bantamlite, Inc.

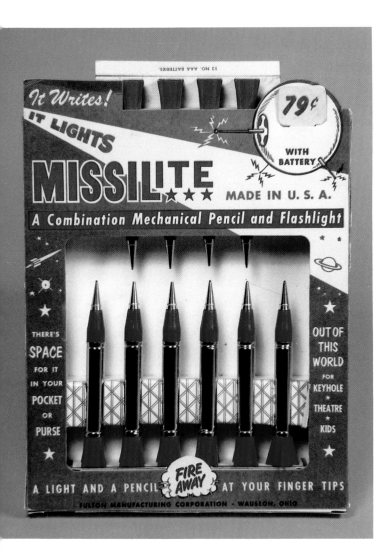

MISSILITE WRITING FLASHLIGHT, ca. 1957. A metal and plastic battery powered flashlight. Note the satellite under the price.

JUNIOR ASTRONAUT FLASHLIGHT, ca. 1962. Paper label on plastic body. These were the most popular types of flashlights for kids in the late 1950s and early 1960s.

TOY ROCKET, ca. 1934. An early painted metal rocket ship. It is about 6″ long and was made in the U.S.A.

STRATOSPHERE GAME, 1934. A game that simulated "A journey through space with a catapult". The game consisted of a target and a metal catapult. The player shot marbles at the multi-level target. It was difficult for the toy manufacturer to capture the essence of space travel. They kept things simple and relied upon the user's imagination.

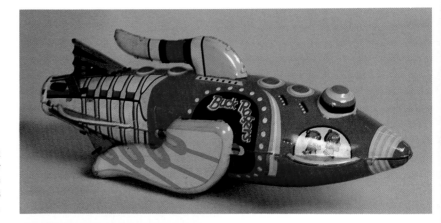

BUCK ROGERS ROCKET, ca. 1934. One of the earliest tin lithographed friction space ships was the Interplanetary Rocket Cruiser. It is 12″ long and was made by Marx in the U.S.A. Buck Rogers, although a fictional character, was many people's introduction to the concept of space travel.

BUCK ROGERS INTERPLANETARY SPACE FLEET, ca. 1934.
Capable carvers could carve creative cruisers with this kit. Instructions
showed how to carve and illustrated the 6 different kits available.

BUCK ROGERS ROCKET, ca. 1939. This tin lithographed friction
vehicle was the Rocket Police Patrol ship. It is 12″ long and was made by
Marx in the U.S.A.

FLASH GORDON PUZZLE, 1951. Three great graphic Flash Gordon puzzles made by Milton Bradley. This was the start of the post war, space toy boom.

ROCKETS AWAY GAME, ca. 1952. The graphics on the box are sensational. The game consists of a large (15″ x 15″) fiberboard target with a periscope type dart dropper (2″ x 4½″ x 9″). The object was to drop a dart on other planets to get points. Made in the U.S.A. by American Metal Specialties.

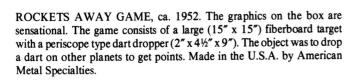

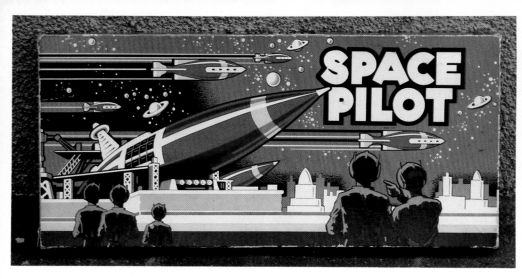

SPACE PILOT GAME, 1952. Space Pilot let you travel to other planets in leaps of millions of miles. The game consists of pegs and a spinner. Advertising says, "Be the first to land on planets with mineral rights worth Billions." A rather unusual selling concept in the toy and game market of the 1950s. Made by Cadaco-Ellis.

CAPTAIN VIDEO SPACE GAME, 1952. Great graphics make this rather ordinary board game into a most exciting one.

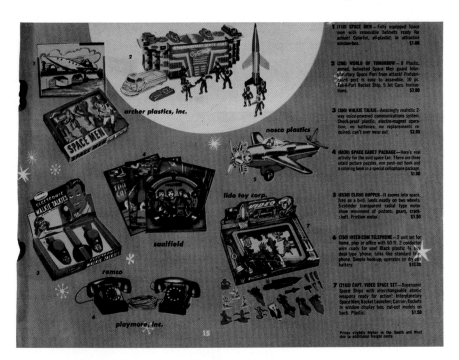

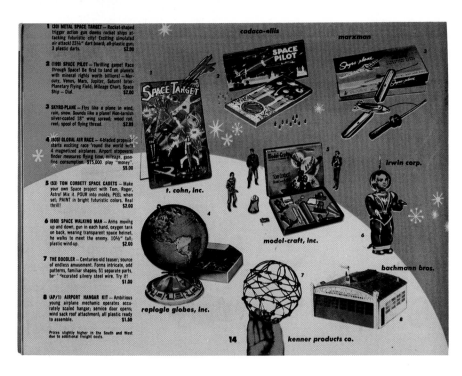

TOY CATALOG, 1952. Two pages of a toy catalog showing the space and technology selections for the Christmas season.

SPACE TARGET GAME, ca. 1952. A lithographed metal target about 24″ tall. It came with a rubber tipped dartgun to shoot down all the jet rockets and missiles. Made by T. Cohn, Inc.

SPACEMAN MOTORCYCLE, ca. 1952. A 6″ long plastic toy with a spaceman holding a ray gun. Probably made in the U.S.A.

SPACE RACE CARD GAME, 1952. Made by ED-U-Cards Mfg. Co. of Long Island City, New York to take advantage of the interest in space. The art work is quite wonderful. The card game was reissued in 1957 after sputnik and again when we landed a man on the moon in 1969. The later issues use the same images on the cards but update the box cover.

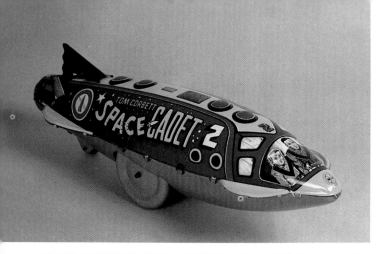

SPACE CADET ROCKET, ca. 1952. Tom Corbett's "Polaris" rocket is 10″ long. A lithographed tin toy made in the U.S.A. by Marx.

SPACE CADET PUZZLE, ca. 1952. Made by Saalfield.

JIGSAW PUZZLE, ca. 1952. "Captain Universe" puzzle. During the 1930s to 1950s, these tray type jigsaw puzzles were made with numerous space action scenes usually aimed at an audience of 4 to 8 year olds. The graphics are bold and colorful.

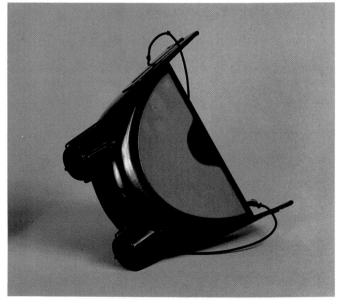

SPACE PATROL BINOCULARS, ca. 1952. Plastic binoculars. The black model was a Ralston Purina Co. Premium. A green version was sold in stores.

SPACESHIPS, ca. 1952. A card of interplanetary spaceships. These simple play sets sold for 10 cents and offered hours of entertainment.

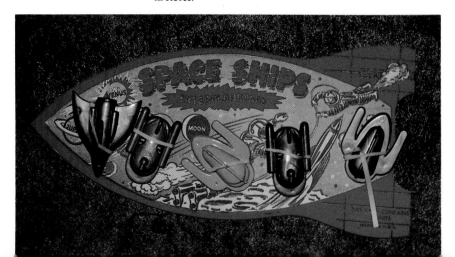

SPACE PATROL PUZZLE, ca. 1952. Made by Milton Bradley.

SPACE PORT, ca. 1953. The Superior Space Port was the space version of the western (cowboy) fort. Lithographed tin shows space scenes and plastic armed (Captain Video) spacemen are attacking. This is believed to be a Captain Video toy whose Captain Video figures are harder to find than the space port.

SPACEMAN ROBOT, 1953. A plastic windup spaceman with ray guns and fishbowl helmet. This was based upon the space adventure show and movies of the early 1950s. 12″ tall.

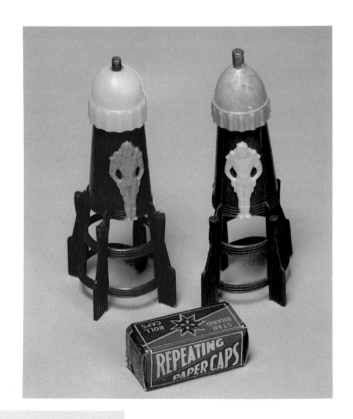

SPACE BAGATELLE GAME, ca. 1954. A paper on wood pinball-type game. Before there were pinball games, they were called bagatelle games.

SATELLITE PORT, ca. 1954. It is a lithographed tin model 12″ long, by Marx.

ROCKET PATROL TARGET GAME, ca. 1954. A lithographed metal target about 14″ tall. Made by American Toy Products, Beverly, Massachusetts.

ROCKET CAP BOMBS, ca. 1954. The top was unscrewed and a cap was loaded into the nose. One replaced the cap and then threw it into the air. When it hit the ground, the cap would explode. Made of plastic.

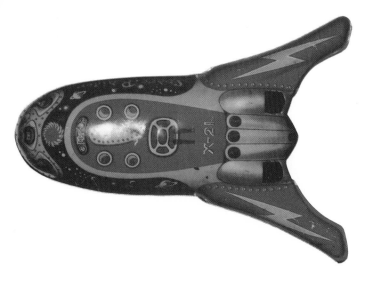

X-21 SPACE RAIDER ROCKET, ca. 1954. Friction, lithographed tin rocket. A nice early space rocket, 10″ long.

DRAFT SCIENTIST DRAFTING SET, ca. 1955. The wonderful graphics tell it all. The young draftsman could design rockets and futuristic cities. Made by Hassenfeld Bros., Central Falls, Rhode Island.

SPACE MAN CAR, ca. 1955. A metal robot in his space car. 6″ long. Made in Japan. A simple friction car with a robot head still allowed the child to imagine that he was controlling a vehicle moving across the moon's surface.

MOON EXPLORER ROBOT, c. 1955. A blue metal and plastic friction w/u robot. 7″ tall. Also made in red. Made in Japan by Yoshiya. Japan was getting its foot in the door by providing services that the American public wanted.

SPACE TOP, ca. 1956. A lithographed tin spinning top with rockets chasing spacemen.

ATOMIC ROCKET, ca. 1956. Lever/friction, lithographed tin rocket. An early space rocket, 6″ long.

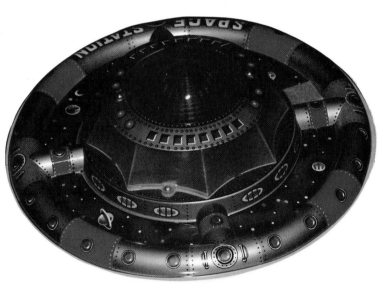

SPACE SPARKING WHEEL, ca. 1956. A Japanese made tin sparking wheel with rockets and stars.

SPACE COMMANDER, ca. 1956. Battery operated, lithographed tin and plastic space station. Lights inside illuminate the crew's activities and the toy moves with a bump and go action. Made in Japan by Nomura.

SPACEMAN WITH LIGHT ON HEAD ROBOT, ca. 1956. A metal and plastic battery operated robot with a remote control. He is ready to explore strange new worlds with all his gear. 7″ tall. Made in Japan for Linemar probably by Daiya.

ROBOT, ca. 1956. A plastic and metal robot. It originally came with a baby robot and was introduced to compete with Ideal's Robert the Robot. 14″ tall. Made by Louis Marx.

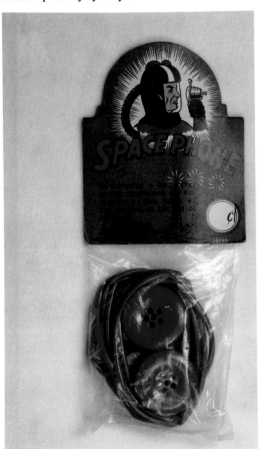

SPACE PHONE, ca. 1956. An early communication device. All you needed was a million miles of thin plastic tubing and you could avoid all of that bulky electronic equipment.

SPACE TANK—V 3, ca. 1956. A windup, flywheel powered, lithographed metal robot driving a tank. It was expected that if man were to land on the moon, he would travel around in a tank like vehicle. The robot is the "Robbie" type. 6″ long. Made in Japan by Yoshiya (KO). This is a rare variation of the blue bodied, battery operated model that is more usually found.

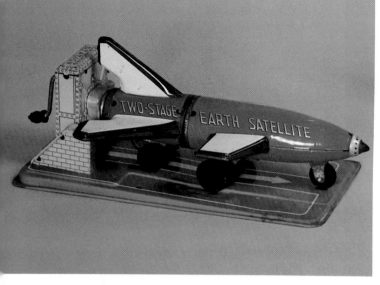

TWO STAGE EARTH SATELLITE, ca. 1956. A two stage lithographed tin rocket with a wind up friction motor. You wound it up, pushed the release lever and away it went. Soon the rear stage would drop away and the front end would continue on its way. Made by LineMar, 10 inches long.

SATELLITE SPIN-AWAY, ca. 1957. Skittle ball with a satellite theme. Made by Consolidated Toy Co.

SPACE PATROL SPINNER, ca. 1956. The rocket spins and points to planets giving the player points. Great fun for long auto trips. The "Space Patrol" name appeared on numerous products, such as this one, that were not related to the show.

JIGGLE PUZZLES, 1957. "Stop the Martians," "Satellite Race," and "Spacemen". Jiggle those little red & white balls into the proper holes and save the planet.

SPACE RACE CARD GAME, 1957. Originally issued in 1952, it was reissued after sputnik and again after we landed a man on the moon in 1969. The reissue updated the box cover.

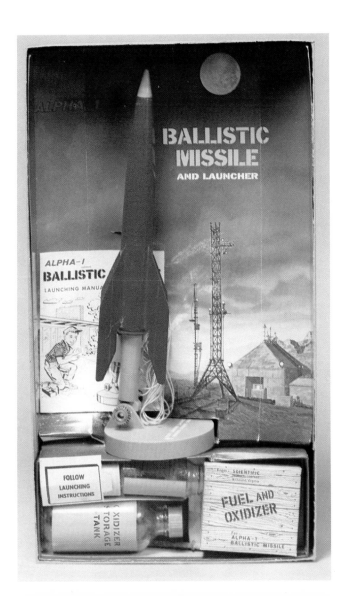

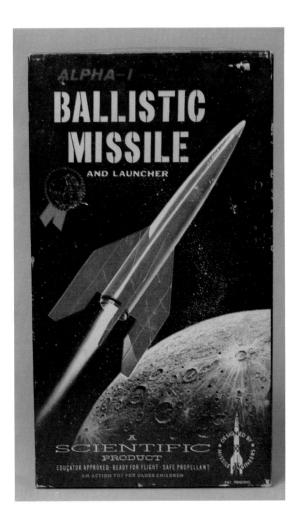

ALPHA-1 BALLISTIC MISSILE AND LAUNCHER, ca. 1957. A baking soda and vinegar powered rocket that was launched by waiting for 8 seconds and pulling the string. If you waited too long, it could explode in a froth of vinegar. Made by Scientific Products Co. of Richmond, Virginia, 3½″ x 6½″ x 12½″.

SATELLITE SHOOT GAME, ca. 1957. An English board game played with chip and cards. Its goal was to see who could shoot down all the rockets and satellites.

MECHANICAL SATELLITE LAND, ca. 1957. A rare, little lithographed tin windup toy with satellite, train and airplane. Made in Japan by K (OHTA), 7″ x 7″.

SPACE SATELLITE, ca. 1957. A mechanical lithographed tin toy with a satellite and a rocket that spun around. Made in West Germany.

MAN MADE SATELLITE, ca. 1957. A mechanical lithographed tin friction toy showing a satellite with a man inside. This has a "we rushed this to market to take advantage of the interest in satellites" look.

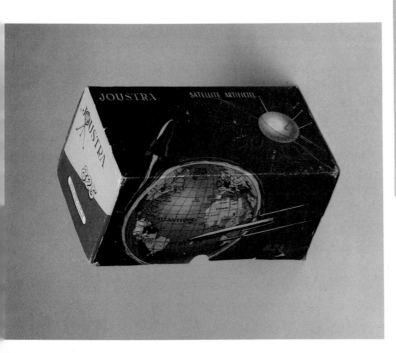

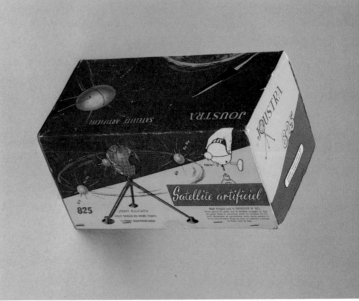

SATELLITE ARTIFICIEL, ca. 1957. This lithographed tin wind-up toy
is probably one of the best visualizations of the holding force of gravity.
The planet turns and as it spins, the satellite pulls out on an elastic cord.
Made in France by Joustra, 13″ high.

X-27 EXPLORER ROBOT, ca. 1958. A blue & red lithographed metal battery powered "trashcan-style" robot. 8.5″ tall. In the 1950s, this was considered a realistic type of spacesuit for venturing out on the moon. Made in Japan by Yonezawa.

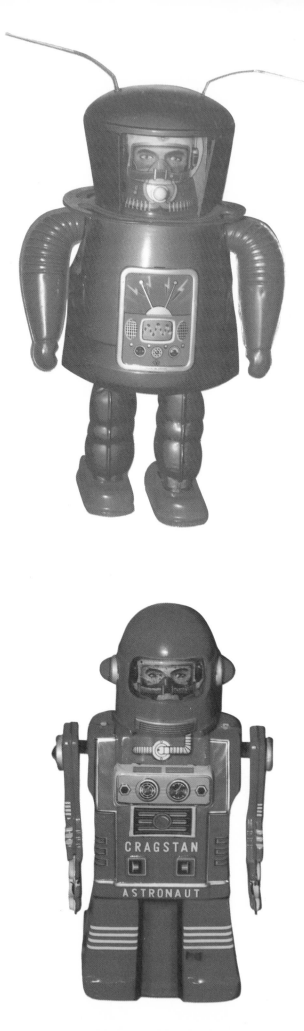

SPACEMAN ROBOT, ca. 1958. A blue & red lithographed metal windup robot. 8″ tall. Made in Japan by Yonezawa.

ASTRONAUT ROBOT, ca. 1958. A lithographed metal battery powered robot. 9.5″ tall. Made in Japan for Cragstan by Yonezawa.

MECHANICAL WALKING SPACEMAN ROBOT, 1958. A metal and plastic battery operated robot. 7″ tall. Made in Japan for Linemar.

SPACE SHIP FLYING SAUCER, ca. 1958. A battery operated lithographed tin saucer. Bump and go action, lights blink in the dome, it stops and a figure rises out of the door. Made in Japan for Cragstan, 10″.

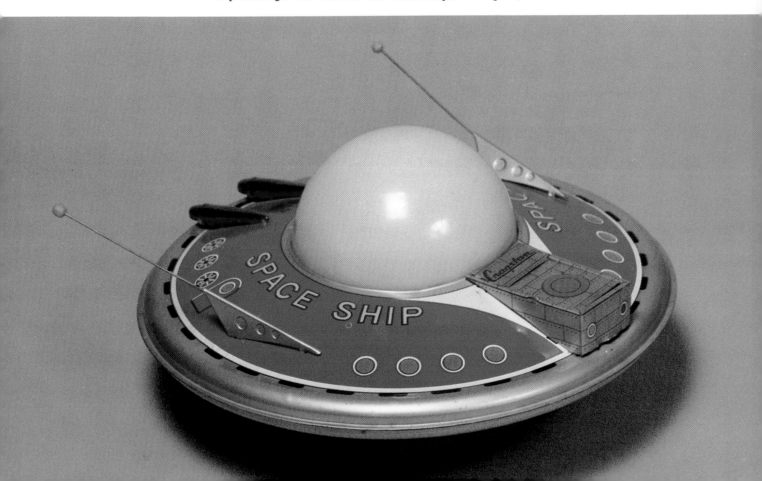

MAN MADE SATELLITE, ca. 1958. More of a piece of space sculpture than a toy. A battery operated lithographed tin toy with a satellite. The earth and satellite turn, the rocket end blinks and the whole thing beeps. This was produced for a very short time. It probably would not hold a child's attention for more than a few minutes. Made in Japan by HOFU, 8½″ long.

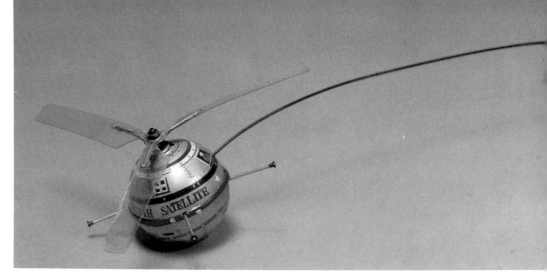

EARTH SATELLITE, ca. 1958. A mechanical lithographed tin toy consisting of a battery holder handle with motor and an attached rotor powered satellite. Made in Japan by Alps, handle is 7½″ long.

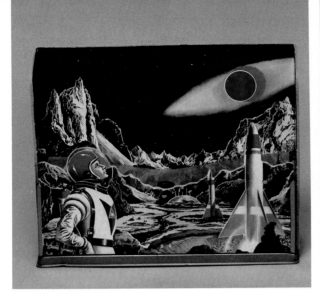

FLOATING SATELLITE TARGET GAME, ca. 1958. A blower supports a styrofoam ball on a column of air and the player shoots rubber tipped darts to knock it down. Made in Japan by S. Horikawa, 6½″ x 9″.

WHISTLING SPACE TOP, ca. 1958. A lithographed toy top with a space theme. Made in Japan for Marx.

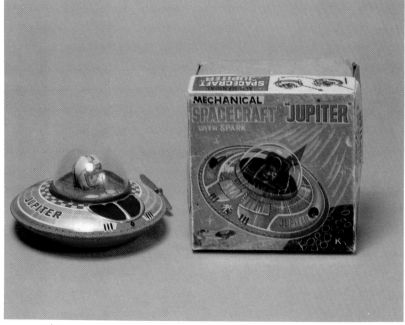

FLOATING SATELLITE TARGET GAME, ca. 1958. A blower supports a styrofoam ball on a column of air above the rocket and the player shoots rubber tipped darts to knock it down. Made in Japan by M.T. (Masudaya), 9″.

MECHANICAL SPACECRAFT "JUPITER" WITH SPARK, ca. 1958. A simple little (3 inches) windup flying saucer that scurried across the kitchen floor and under your formica kitchen table. Made in Japan by K (OHTA).

SATELLITE IN ORBIT, ca. 1958. A great battery operated toy. The world turns and a styrofoam ball floats on the stream of air created by the little blower. Made in Japan For Cragstan by S. Horikawa, 9 inches tall.

SATELLITE, ca. 1958. The magnetic sputnik revolves around the planet on the rails and a battery in the main body makes it light up. Made in Japan, 10″. This is a rare model with the earth being lithographed tin. There is a similar toy with a plastic earth in the center.

SPACE REFUEL STATION, ca. 1959. A metal and plastic battery operated space station with lots of action—bump & go action with spinning antenna and blinking lights, it rises up and revolves. 14″ tall. Made in Japan by WACO.

SPACE STATION, ca. 1959. A metal and plastic battery operated space station with a great deal of attention to detail—bump & go action with spinning antenna and blinking lights. 14″ across. Made in Japan by Horikawa.

PIONEER 3 STAGE ROCKET, ca. 1959. Spring propelled, lithographed tin rocket. As it goes up, a nose cone dart-like piece is released. Made in Japan for Kramer, 10″ high.

PEDAL CAR, ca. 1960. Murray "Atomic Missile" pedal car.

ASTRONAUT ROBOT, ca. 1960. A metal and plastic battery operated robot. 13″ tall. Made in Japan for Cragstan possibly by Daiya.

ASTRONAUT ROBOT, ca. 1960. A metal and plastic battery operated robot. 12″ tall. This model was also available in red. Made in Japan for Rosko.

FLYING SAUCER, ca. 1960. Tin bump and go battery operated saucer with a lithographed spaceman inside. Made in Japan by K.O. (Yoshiya) for Cragstan.

MECHANICAL SATELLITE FLEET, ca. 1961. A great little windup toy with flying saucers that resembled a momma duck followed by three babies. Created in response to the growing interest in flying saucers. Made in Japan by T.P.S., length 13 inches long.

ROCKET RIDE, ca. 1960. A windup lithographed tin toy based upon the real models that adorned major amusement parks in the 1950s. (See the illustration of the 1948 Chambers rocket ride.) Stands about 24 inches tall.

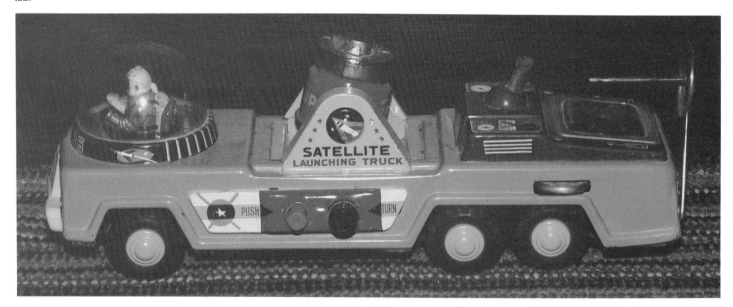

SATELLITE LAUNCHING TRUCK, ca. 1960. A metal vehicle. 12″ long. The spaceman drives to a likely looking spot on the moon and then launches his satellites. Made in Japan by Yonezawa.

MOON EXPLORER, ca. 1961. A metal and plastic battery operated moon vehicle with great attention to detail—walking action with spinning antenna and blinking lights. 10″ long. Made in Japan by Yonezawa.

MECHANICAL WALKING SPACE PATROL ROBOT, ca. 1961. A metal and plastic w/u robot. 6″ tall. A typical robot with a space suit and astronaut head. He looks very much like the astronauts from the Mercury space project. Note the unauthorized use of the NASA logo. Made in Japan.

PLANET Y SPACE STATION, ca. 1962. Tin bump and go battery operated toy with a plastic spaceman inside. Space stations enjoyed brief spurts of interest. The idea of the station was very appealing but turning it into a toy left something to be desired. Made in Japan by T. Nomura.

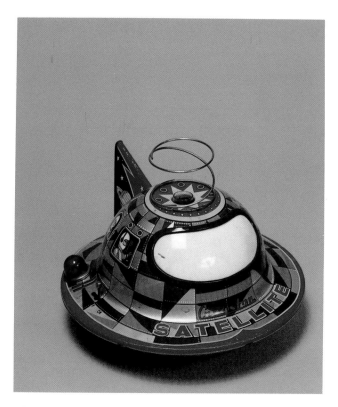

SATELLITE FLYING SAUCER, ca. 1962. Lithographed tin bump and go saucer that contained a blower that would keep a styrofoam ball or little spaceman suspended in air as it moved. The clock work mechanism made the blower blow faster and then more slowly so the ball moves up and down. Made in Japan by Masudaya. 8 inches.

CAPSULE "MERCURY", ca. 1962. Friction powered red lithographed tin capsule marked "Friendship 7". This was the name of John Glenn's capsule. The astronaut inside revolves as it moves and sparks shoot from the tail. It is 10 inches long and made in Japan by Horikowa.

MERCURY CAPSULE, ca. 1962. Another version of the "Friendship 7" capsule in yellow. Also 10 inches long and made by Horikowa.

ASTRONAUT GAME, ca. 1962. A nice simple card game to pass the time of day.

TWO-STAGE ROCKET LAUNCHING PAD, ca. 1962. Battery operated, the countdown is set in motion and the lights flash, the man's arms move and the panel beeps. Finally the rocket shoots out the tube. The design would never pass child safety standards today. Made in Japan by T. Nomura, 4″ x 7″ x 8″.

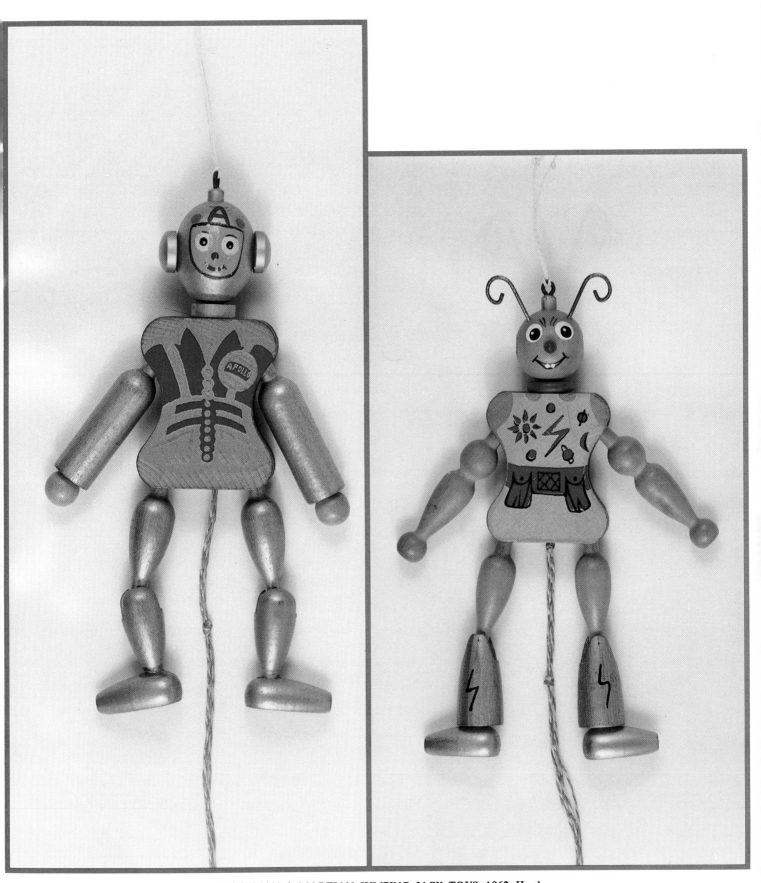

SPACEMAN & MARTIAN JUMPING JACK TOYS, 1962. Hand painted wooden pull toys produced by Gregor Creations, Inc, New Orleans, Louisiana. 8 inches high. Unusual since the design gives one a 1950s, UFO period "feel" but the pieces are copyrighted 1962.

SATELLITE MONITOR, ca. 1963. A metal and plastic battery operated satellite tracker with spinning antenna and blinking lights. 9" long. Made in Japan by Yonezawa for Cragstan.

FRICTION MOON ROCKET, ca. 1964. 16" tall, lithographed tin rocket that was pushed forward. When it struck something, it stood up and the stairway descended to expose an astronaut. Made in Japan by Matsuta.

CAPSULE 6, ca. 1964. Battery operated, handsomely lithographed tin and plastic capsule. The nose cone lights and the spaceman rocks, holding a camera, as the capsule moves on the floor in typical bump and go action. Made in Japan by TM (Masudaya), 13" long.

MOON ROVER, ca. 1964. Battery operated, lithographed tin and plastic vehicle. The lights blink as it moves on the floor. Made in Japan by TM (Masudaya), 10"long.

NASA COUNTDOWN and RAK 110 ROCKET, ca. 1965. The screen lights up and two space capsules appear. With radar dish spinning and use of the controls, you can dock the two ships on your NASA Countdown. NASA Countdown is unmarked and most likely was made in Japan. The Rak 1100 is a product of West Germany. It would not meet today's consumer product safety standards since the rocket is loaded onto a powerful spring and when the gantry is removed, the rocket can go in unexpected directions. West German space toys vary in quality and often show good concepts but poor execution.

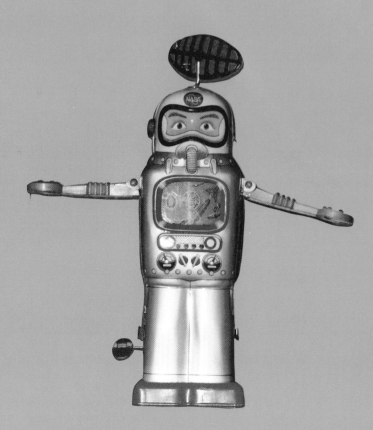

TELEVISION SPACEMAN ROBOT, ca. 1965. A metal and plastic windup robot. 6″ tall. As he walked, the "television" in his chest turned with changing scenes. Made in Japan by Alps.

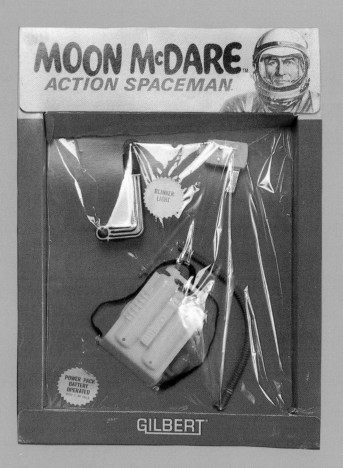

MOON McDARE SPACEMAN EQUIPMENT, 1965. Accessories to compliment the Moon McDare Spaceman doll. Made by Gilbert.

MOON McDARE SPACEMAN EQUIPMENT, 1966. Accessories to compliment the Moon McDare Spaceman doll. Made by Gilbert.

SOLAR CONQUEST GAME, 1966. This space game has you travelling
from earth to the moon and the planets. Made by Tech Enterprises.

BULLET SKYLINER, ca. 1966. A sort of monorail set adapted for use
on the moon. Made in Japan by Yonezawa.

SPACE CAPSULE, ca. 1967. Battery operated, lithographed tin and plastic. The nose cone lights, the doors open, the whole thing squeals and the spaceman aims his camera while looking at a television screen. The capsule moves on the floor in typical bump and go action. Made in Japan by S.H. (Horikawa) 9½″ long.

COUNTDOWN GAME, ca. 1967. Now anyone could make the flight into space aboard a rocket from the safety of their home. Made by E.S. Lowe/Mondotoy.

APOLLO—Z, ca. 1967. Battery operated, lithographed tin and plastic. The tail and nose cone light and the radar on the top of the capsule turns as the capsule moves on the floor in typical bump and go action, then it stops, the front lifts up and the front end of the capsule moves away from the body. It then reverses and moves along again. Made in Japan, 12″ long.

SPACECRAFT APOLLO, ca. 1967. Battery operated, lithographed tin and plastic. The tail and nose cone light and the radar on the top of the capsule turns as the capsule moves on the floor in typical bump and go action. Made in Japan by Alps, 9″ long.

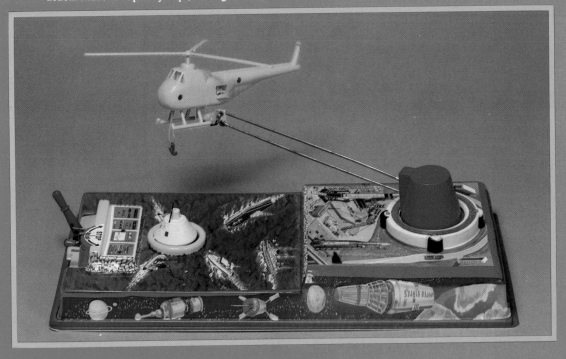

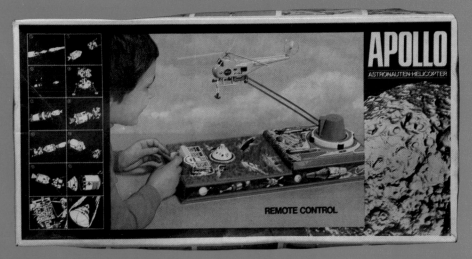

APOLLO ASTRONAUTEN-HELI-COPTER, ca. 1968. Battery and hand operated vacu-formed plastic and metal. The Helicopter goes round and round. The operator makes it hover over the capsule while trying to snag it with the hook. A miniature astronaut on a stretcher can be loaded into the side of the helicopter to wisked off to safety. Makes you wonder why they put the astronaut on a stretcher? Made in West Germany, 9″ x 18″ long.

LUNAR CAPTAIN, ca. 1969. A battery operated plastic and metal toy
of amazing complexity. The whole thing starts out in a horizontal position
and starts blinking. It then stands itself upright and the space capsule
separates from the Lunar Lander, revolves and then redocks. It then sets
itself down horizontally and begins to move again. Made in Japan by T.
Nomura, 12″ tall.

EAGLE LUNAR MODULE, ca. 1969. Battery operated lithographed
tin and plastic. Bump and go action with the radar antenna turning and
lights flashing. It stops, a metal door opens and an astronaut in a silver suit
appears next to the ladder. Made in Japan by DSK (Daishin), 7″ x 7″ x
9½″.

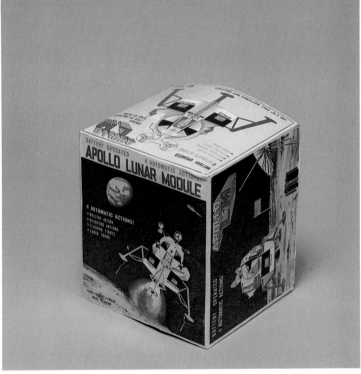

EAGLE LUNAR MODULE, ca. 1969. Battery operated lithographed
tin and plastic. Bump and go action with the radar antenna turning and
lights flashing. Made in Japan by DSK (Daishin), 5″x 5″x 7½″.

TRANQUILITY BASE EAGLE LUNAR MODULE, ca. 1969. A
Revell model kit and box. This is the most accurate model kit of the
lander.

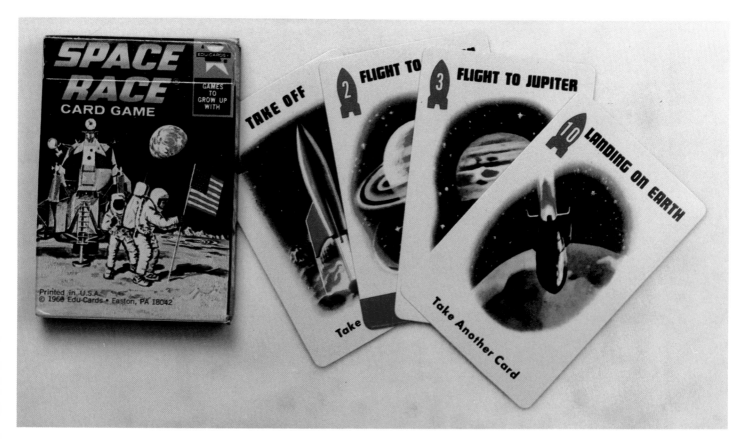

SPACE RACE CARD GAME, 1969. The card game originally issued in
1952, was reissued in 1957 after Sputnik was launched and again after we
landed a man on the moon in 1969.

JIGSAW PUZZLE, ca. 1969. The graphics are unusual. This illustration emphasizes the men, not the mission. Most items seemed to pay homage to the men but emphasized the mission.

ORBIT YO-YO, 1969. Every simple toy was given a space theme to boost sales when it looked like we were about to land a man on the moon.

SATURN ROCKET MODEL, 1969. Model rocket enthusiasts could launch their own moon rocket with this large scale model. Three or four small "B-8-4" rocket engines would lift this monster a few hundred feet into the air. Parachutes would bring it back softly.

SPACE SCOOTER, 1970. A battery operated model of an astronaut on his space scooter. Made in Japan by Masudaya, length 9 inches long.

APOLLO MOON EXPLORING TOY, 1970. Made by Imperial Toy Corp in Los Angeles, California. Using the space theme and the "collect all 12 sets", toy manufacturers figured to cash in on the space craze.

SPACE SHUTTLE 101 GAME, ca. 1978. "Command actual missions into outer space for the benefit of mankind".

APOLLO 12 SPACE CAPSULE TOY, 1970. A small friction powered model of the Apollo capsule. Made in Japan by T.T., length 4 inches long.

ROBOT, 1977. Would a robot be Man's eyes and ears as he set foot on distant planets? This 5″ tall fellow, modelled on the Lost in Space robot did not do much. He just lit up and could be pushed along.

R2D2, 1978. This little 8″ tall fellow, from the movie *Star Wars*, was radio controlled. He would turn his head and move in the direction that the head aimed. Star Wars toys, action figures and other merchandise were created in immense numbers. This item was rather expensive and fewer were sold. Made by Kenner.

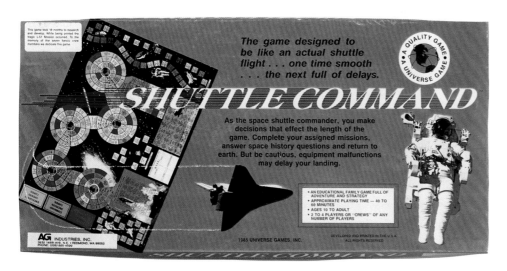

SHUTTLE COMMAND GAME, ca. 1985. Another attempt to translate the thrill of the space flight into a board game. Made by Universe Games/A.G. Industries, Redmond, Washington.

SPACE COLLECTIBLES THROUGH THE YEARS

A SIGNAL FROM MARS, ca. 1901. A sheet music cover. Sheet music often sold by appealing to the popular themes of the day. It was about this time that Professor Percival Lowell, a noted astronomer, announced that he was receiving signals from the planet Mars. He was watching them and they were watching us. Martians were trying to contact us.

A TRIP TO THE MOON, ca. 1910. The cover of a piece of sheet music. Halley's comet reappeared in 1910. Scientists predicted that the earth would pass through the comet's tail. In 1908 Professor Morehouse discovered that the tail of a comet contained cyanogen gas. Cyanogen gas, when mixed with our atmosphere would create prussic acid a deadly poison. Ignorant people believed that there should be some attempt made to flee from the earth into space. Numerous farfetched ideas were conceived, among which was to take an airship to the moon. Clifford V. Baker set the idea to music.

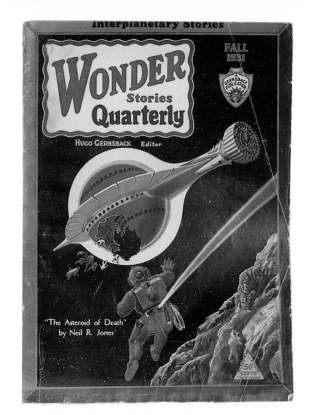

WONDER STORIES QUARTERLY, 1931. Hugo Gernsback was one of the earliest editors of science fiction magazines. Unfortunately these magazines were printed on a cheap paper that crumbles over time. They are fondly referred to as "Pulps".

POSTCARDS, 1910. In 1910, Halley's comet appeared as predicted. When the public found out that the comet's tail contained poisonous gas, some people panicked and wanted to flee the earth for the moon. These postcards were spoofs on the ideas that people dreamed up to get to the moon. One is from Germany and the others are from France. "Weltuntergang" and "Fin Du Monde" mean "End of the World". Note the photographer offering to take pictures of the end of the world spectacle for 20 francs, "Payable in advance".

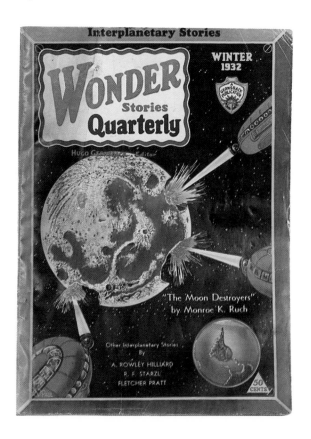

WONDER STORIES QUARTERLY, 1932. The pulps are usually collected for their great cover art and early stories by later famous authors.

BUCK ROGERS COMICS, October 1, 1933. Buck and Wilma on Mars in the Sunday funnies.

WILMA DEERING, ca. 1934. A paper mask of Buck Roger's companion. Buck Rogers was the most popular spaceman of his day. His constant flights to explore distant planets and keep the earth safe from its enemies caught the public's attention and apparently sold shoes. The back is marked "I always buy my Tri-Tan or Powhattan shoes at...".

BUCK ROGERS MAP OF THE SOLAR SYSTEM, ca. 1933. A Cocomalt premium with excellent graphics.

BUCK ROGERS PIN, ca. 1935. A Buck Rogers club member's full color button offered as a premium. It is 1 inch wide.

BLAZING STAR & SHOOTING STAR BRAND FIRECRACKERS, ca. 1940. Rare labels with a rocket. The dream of a boy riding a rocket into space is captured in the art work of these labels.

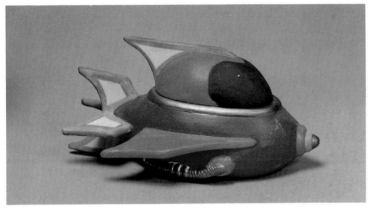

ROCKET SHIP, design ca. 1935. A great Flash Gordon type rocket made of ceramic. This is a modern made desk accessory that captures the flair and color of the early space adventure rockets.

BUCK ROGERS BADGES, ca. 1936. A Spaceship Commander whistle badge and a 1934 Cream of Wheat premium Solar Scouts Member badge. Both about 1.5 inches wide.

BUCK ROGERS ADVERTISEMENT, 1940. Buck and Wilma, when not rocketing off into space, act as spokesmen for several products.

FLYING STAR AND FLYING FAIRY BRAND FIRECRACKERS, ca. 1930 & ca. 1949. Two very rare labels. Flying Star is the typical Buck Rogers image of a mysterious planet and Flying Fairy is an early accurate depiction of a rocket ship.

INTERPLANETARY CHART, ca. 1946. Paper on board shows all the distances to other planets.

ROCKET SHIP GALILEO, 1947. A whole collecting field in itself, space books provide information and entertainment. This one was written for teenage readers about three boys who accompany a scientist on the first journey to the moon. Author is Robert A. Heinlein.

AMAZING STORIES, 1948. Amazing Stories was the pulp that introduced Buck Rogers.

STARTLING STORIES, 1947. A popular subject of the cover art of the pulps was the beautiful girl with a ray gun, often being rescued by the handsome space hero.

ROCKET TO THE MOON PAINTING, ca. 1947. A black and white painting of a rocket to the moon. This may have been an illustration for one of the science fiction magazines. Artist unknown. Size of the art work is about 7″ x 8″.

ROCKET AMUSEMENT PARK RIDE, ca. 1948. R.E. Chambers of Beaver Falls, Pennsylvania made these 21 foot long rocket cars for amusement parks. This 850 pound rocket that seated 8 people could replace the then current circle swings and seaplanes. Chamber's advertising said, "Of bright, gleaming, polished stainless steel, the cars are always clean and attractive and make a very spectacular appearance as they flash in the sunlight or reflect surrounding night lights." Look for an illustration of the toy version of this ride further back in the book. The rider wears his Tom Corbett Space Cadet hat made about 1954.

ASTRONAUT AND ROCKET BRAND FIRECRACKERS, 1950s. Larger than one and a half inches, these were the firecrackers that could blow off a finger if they were held too long. Most kids lit them and ran.

MISSILE MAIL ASHTRAY, 1950s. The original missile mail was a 1930/1940s idea. Well, it seemed like a good idea at first—shooting the mail from here to there. Then the postal authorities decided that using guided missiles to send the mail was not a realistic idea. Just imagine, "Honey, leave the window open, I'm expecting a special delivery letter today".

AMAZING STORIES, 1951. Again an illustration of a beautiful girl with a ray gun.

FANTASTIC STORY, 1951. A beautiful blonde with a ray gun. The cover art gives a clue as to who was reading these publications.

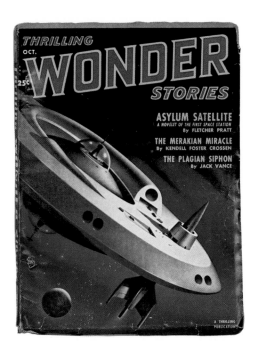

THRILLING WONDER STORIES, 1951. The cover art shows a rocket and space station.

SPACE CADET BADGE, ca. 1951. A high quality metal badge with the Space Cadet insignia. Two inches wide.

RECORD, ca. 1951. "Tom Corbett Space Cadet Song and March," 45 rpm record with the orchestra conducted by Mitchell Miller, Rockhill Productions, The back side shows the space academy, the characters and a space port.

ROCKET SIGN, ca. 1951. Oldsmobile introduced their Delta 88 in the late 1940s. It was compared to a rocket and their advertising and service signs used this image.

ROCKET CARNIVAL RIDE, ca. 1952. Some amusement parks built rocket rides from discarded airplane fuel tanks. The rocket rider wears a "Satellite Explorer" aluminum space helmet made about 1954.

MOVIE POSTER, 1953. Commando Cody, Sky Marshall of the Universe, in "Nightmare Typhoon". by Republic Pictures. Commando Cody was one of the "Buck Rogers" of the 1950s that protected the earth from attackers and aliens.

SPACE PATROL PLATES, ca. 1952. Kids love 'em.

THE ROCKET, ca. 1953. An amusement ride 6 feet long and weighing about 500 lbs. It was activated by dropping a dime in the coin box. One control makes the rocket nose upward at a 45 degree angle, another control makes the whole rocket drop and the third slowly drops the rocket and then it raises up and down. The lights flash on and off. It really is quite an adventure compared to today's simple kiddie rides.

TOM CORBETT SPACE CADET BELT, ca. 1953. This was perfect belt upon which to hang a holster and space gun.

SPACE COMMANDER COSTUME, ca. 1953. This heavy canvas outfit may have been for a space character in Great Britain.

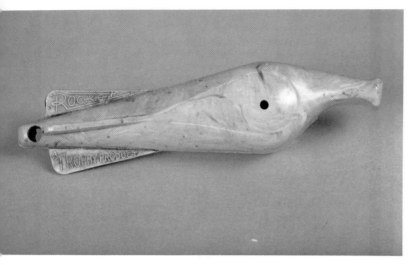

ROCKET WHISTLE, ca. 1953. The whistle to call all space rangers to attention. A good example of the influence of space age styling on everyday items. Made in the U.S. by Trophy Mfg.

SPACESHIP WHISTLE RING, ca. 1953. A metal ring made in Japan for Preview Productions, Inc.

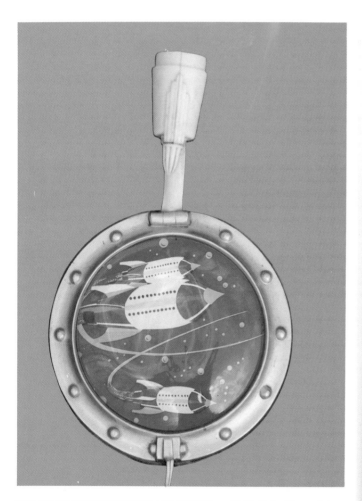

SPACESHIP PORTHOLE WALL LIGHT, ca. 1953. Probably made to go next to the young space ranger's bed so that he could look out into space and dream. Molded plastic 10″ across. It may have had a space scene lamp shade.

FROM THE EARTH TO THE MOON BY JULES VERNE, March, 1953. A Classics Illustrated comic book. Classics Illustrated reprinted classic books—this one is from the late 1860s—in a comic book format. It was possibly the only access to the classics that many children would be exposed to.

FLYING SAUCER CARNIVAL RIDE, ca. 1954. "I then saw this horrible looking alien swoop down in a tiny flying saucer and I became frozen by his ray gun."

SPACE RANGER BADGE, ca. 1954. A metal badge with the Rocky Jones, Space Ranger insignia. Three inches wide.

FLYING SAUCERS MATCHBOX LABEL, ca. 1954. Produced during the UFO craze.

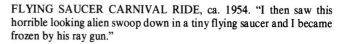

SUNGLASSES, ca. 1954. These Rocket sunglasses sport tiny rockets and a spaceman. Made by Fosta. Awfully cute, don't you think?

SPACE RANGER, ROCKY JONES WRISTWATCH, 1954. Rocky stands poised before his rocket, helmet in hand, ready to save earth on a moment's notice.

SPACE SHIP KNIFE, ca. 1954. Two blade pocket knife with a Morse code chart on the reverse side.

ROCKET KNIFE, ca. 1954. Two blade pocket knife with the words "Jet Jacknife".

NOVELTIES, ca. 1954. Small plastic charms that might have been available in gumball machines. Ray guns, rockets and spacemen were the main themes.

SPACE PILOT TELESCOPE, ca. 1954. 4½ inches tall, inexpensively decorated tin telescope with a rocket and a spaceman.

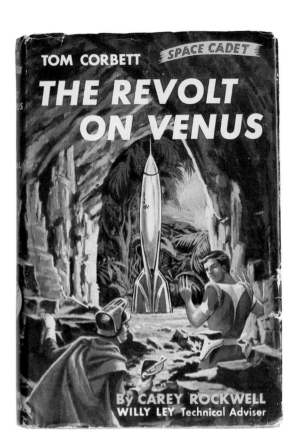

THE REVOLT ON VENUS, 1954. A Tom Corbett, "Space Cadet" adventure book. The year is 2353. Tom and his two cadet buddies go hunting on Venus and stumble upon a renegade force preparing to attack earth.

SPACE GOGGLES, ca. 1954. The earliest spacemen wore goggles on their flights. The "Magic" aspect was that these were silvered on the outside so no one could see your eyes.

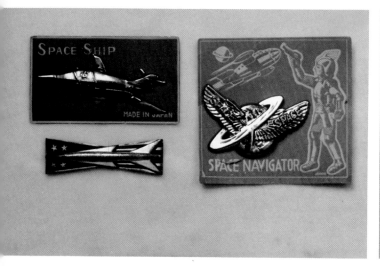

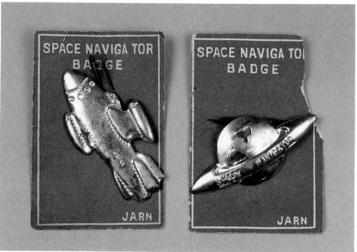

SPACE SHIP, ROCKET & SPACE NAVIGATOR PINS, ca. 1954. The two pins on the cards are made of tin while the rocket pin is sterling silver. The Space Navigator pin is nearly identical to a Rocky Jones Space Ranger pin.

SPACE NAVIGATOR BADGES, ca. 1954. The two Japanese made pins are tin. The Space Navigator name was a Rocky Jones, Space Ranger logo. Note that they misspelled "Japan" at the bottom of each card.

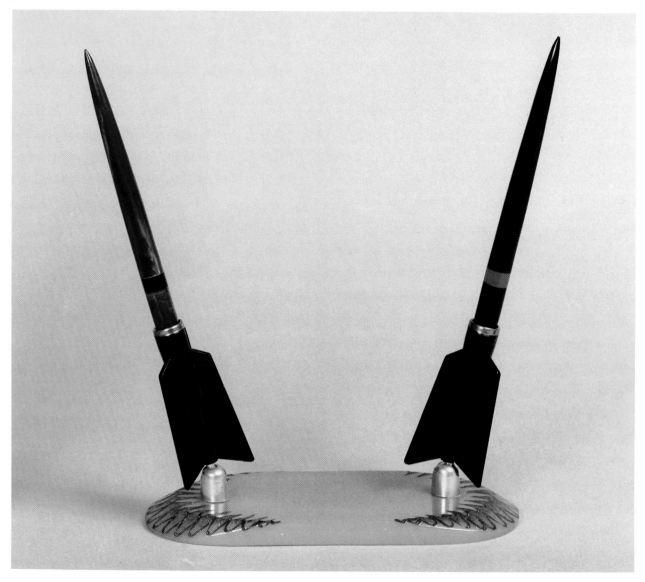

ROCKET PEN & PENCIL, ca. 1954. A Bakelite pen and pencil set.

SPACE CHARM BRACELET, ca. 1955. A Space Kid-Ette charm bracelet with a compass, spaceman, rocket, lovely space girl sidekick, robot and ray gun. The piece was probably made in Japan and carded in the U.S.

POCKET WATCH, ca. 1955. A flying saucer image with a spaceman.

COUNTDOWN MUG, ca. 1955. Mugs are difficult to date but the graphics are in the mid-1950s style.

GUM BALL MACHINE, ca. 1955. A 22 inch tall "Rocket" counter-top gum ball machine.

SPACE TRANSFER PICTURES, ca. 1955. We called them tattoos. They only cost a quarter and there were enough sheets of tattoos to split them with friends and still have plenty. You licked your arm and held it in place for 15 seconds or so. The better method was to wet your arm, stick on the tattoo, and hold it down with a wet washcloth. They washed off with water. Arms would be covered with rockets, space scenes and spacemen. Mothers hated these things.

MARTIAN STATUE, ca. 1955. 10 inch tall wooden statue of a Martian, carved and signed by Ray Monroe. Was this one of a kind or did Ray Monroe carve these as a business?

SPACE HANDKERCHIEF, ca. 1955. Great images of young astronauts frolicking on an alien planet.

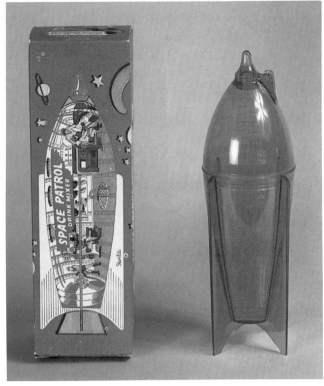

SPACE PATROL DRINK MIXER, ca. 1955. Imagine all the little space patrollers making whiskey sours. Actually those were the days that you made your own chocolate milk with Bosco or Ovaltine or Nestles. Made by United Plastic Corp. of Fitchburg, Mass.

CHOCOLATE SYRUP CONTAINERS, ca. 1955. Since the kids were drinking chocolate milk, some manufacturers sold their syrup in bottles shaped like spacemen or robots.

DRINK MIXERS, ca. 1955. With all the children drinking chocolate mixes with their milk, drink mixers were great items for mom.

SPACE SCOUTS STARDUST BUBBLE BATH, ca. 1955. Bubble bath powder in packets with rockets and planets on them. 20 packets per box. More than thirty years later I can still remember the smell of the soap. It was real industrial stuff that irritated your nose and left a great grey bathtub ring.

ROCKET COAT TREE, ca. 1955. A nicely made coat tree in the shape of a rocket that was possibly made by a shop handy dad for his son's room.

SPACE NAVIGATOR PIN & COMPASS, ca. 1955.

CEILING LIGHT COVER, ca. 1955. Ceiling light covers came in many designs. This spaceship filled design must have added a feeling of space adventure to some child's room.

ROCKET SHIRT, ca. 1955. A Hawaiian style shirt with rockets and spacemen all in vibrant colors.

FLYING SAUCER AND SPACEMAN BOOK ENDS, ca. 1955. Wonderful whimsical handmade wooden bookends with a group of good reading books -*UFOs For The Millions* by H.V. Chambers (1967), *Beyond Mars* by Nephew & Chester (1960), *First On The Moon* by Armstrong, Collins & Aldrin (1970) and *First Men To The Moon* by Werner Von Braun (1960).

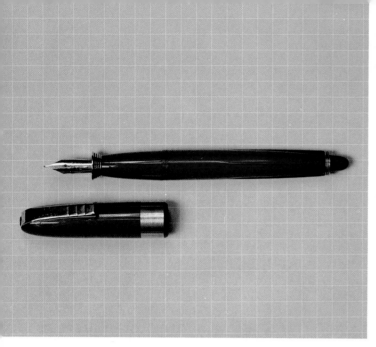

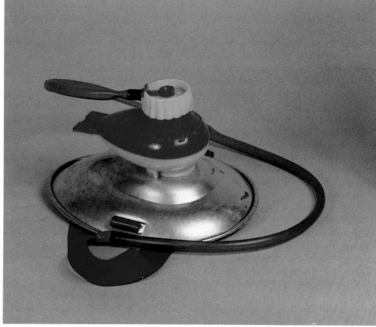

FOUNTAIN PEN, 1956. The Stratford Magnetic fountain pen was made in the mid-1950s. It had small rocket fins at the bottom, a streamlined cap, a "Jetsons" like collar and used magnetism to hold the cap on either end.

SPACE BUBBLE HAT, ca. 1956. "Bubble-O-Bill" space hat that allowed you to blow bubbles while wearing it. Rather unusual.

SPACEMASTER TOY CHEST, ca. 1956. A lithographed tin toy chest with rockets, spacewomen and spacemen. 18″ x 36″ x 18″.

ACE BRAVE—SPACEMASTER BOOK, ca. 1956. English pop-up book. Ace is bound for Mars, the most popular destination in the 1950s.

MOVIE POSTER, 1956. "Satellite in the Sky". "The never told story of life on the roof of the earth," by Warner Bros. The idea of putting a satellite in orbit above earth was about to become a reality.

ROCKET SALT & PEPPER SHAKERS, ca. 1956. Salt & pepper shakers were made in every shape and size. Moon rockets were a natural for salt & pepper.

117

ROCKET SALT & PEPPER SHAKERS, ca. 1957. Salt & pepper shakers made in the shape of rockets.

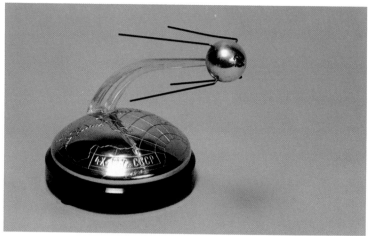

SATURN LAMP, ca. 1956. 11 inch tall lamp in the shape of Saturn. This was perfect for the youngster's room decorated with space ships and planets.

SPUTNIK MUSIC BOX DESK ORNAMENT, 1957. A Soviet-made model of the sputnik that played the U.S.S.R. anthem and then made the ping, ping, ping sound of the sputnik. These were available in Russian stores and at the Brussels World's Fair in 1958 and are usually found, if at all, in Europe.

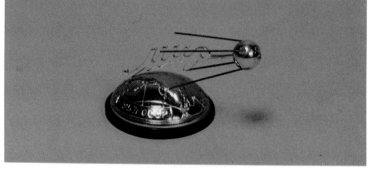

TO MARS SALT & PEPPER SHAKERS, ca. 1956. Our barely dressed traveler has his bag packed and is about to board his rocket to Mars.

SPUTNIK DESK ORNAMENT, 1957. A tiny Soviet-made model of the sputnik soaring above the earth. Note the letters, CCCP, in the support. This was the Russian spelling of U.S.S.R.

SPUTNIK MUSIC BOX DESK CLOCK, 1957. A Soviet-made model of the sputnik that played the "ping, ping, ping" sound of the sputnik with a Swiss-made watch set into the sputnik's nose.

ROCKET MUD FLAP, 1957. A bicycle mud flap with a rocket. These were the days when your "wheels" could be turned into a spaceship.

POSTCARDS, 1957. These Russian postcards honor the first flight of the sputnik satellite. These are early and difficult to find.

THERMOS FLASK, ca. 1957. Shows space stations, rockets and planets. Made by Thermos for a space lunchbox.

119

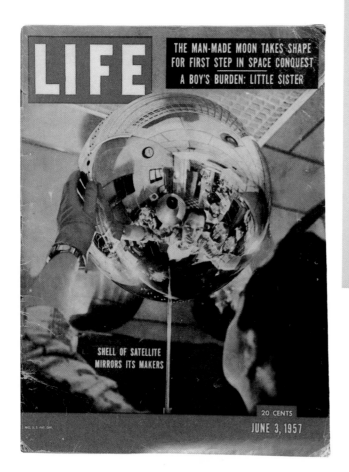

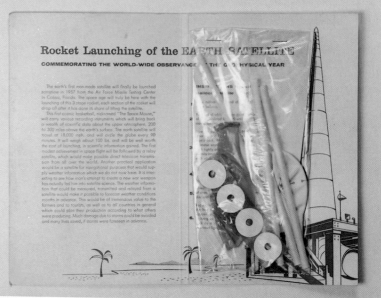

EARTH SATELLITE, 1957. This enterprising manufacturer created a replica satellite made with a balloon within a balloon and 4 straws sticking out to represent the antennas.

LIFE MAGAZINE, June 3,1957. It was expected that the United States would launch its satellite in 1957 to commemorate the world-wide observance of the Geophysical year. This issue covered the building of the satellite. *Life Magazine* made a deal with NASA for the exclusive personal coverage of the Mercury 7 astronauts. *Life* was a good source of information and photographs about the space program.

SPACE BLANKET, ca. 1958. A colorful blanket with loads of space action from launch to orbiting satellite.

SPACE CLOCK, ca. 1958. Made in Scotland, this interesting Westclox made space clock is a fun accessory for a boy's room.

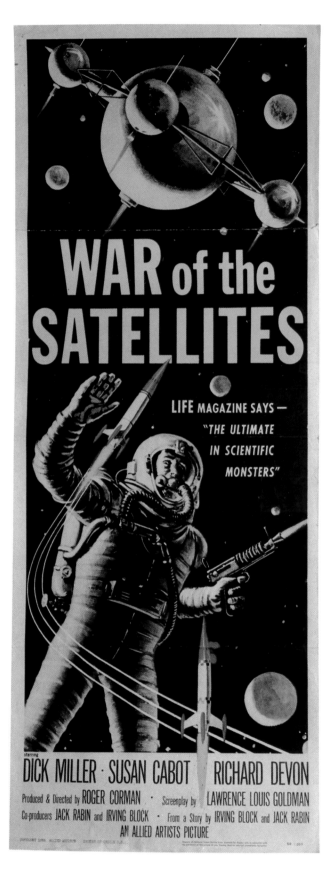

SPUTNIK PUZZLE, 1958. The puzzle is a standard wood puzzle of the 1950s with a Sputnik label attached to take advantage of the new interest in sputniks. Made in Japan.

MOVIE POSTER, 1958. "War of the Satellites". "The ultimate in scientific monsters" by Allied Artists. Now that satellites were actually in space, it was appropriate for them to do battle.

MEDALLION FOR EXPLORER 1, 1958. 1.25″ in diameter. Explorer 1 was our first satellite launched into space. The difficulty with dating a medal such as this is that it may not have been issued in 1958. It may be part of a later set, issued at anytime.

SATELLITE TIN WHISTLE, 1958. The whistle is lithographed tin with great art work. Made in Japan.

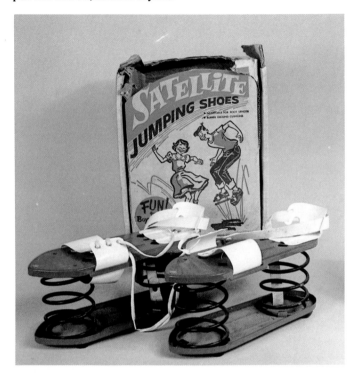

SATELLITE JUMPING SHOES, 1958. Don't jump too high or you may find yourself in orbit. This is a good example of a manufacturer attaching a space theme to an existing product to generate extra sales created by the interest in space.

SATELLITE BABY MOON PENCIL SHARPENERS, 1958. A lithographed tin pencil sharpener in the shape of a satellite made by Sterling Plastic of Union, New Jersey.

SPUTNIK POCKET KNIFE, 1958. A flasher image of the Sputnik changes to a reclining naked lady with the words "Out of this World".

SPUTNIK SODA GLASS AND HOLDER, 1958. At the 1958 Brussels World's Fair, this stylized sputnik was the logo of the U.S.S.R.

U.S. SPACE STATION, 1958. A nice space station model kit by Lindberg Products. The design was realistic with solar gatherers, landing area, rockets to turn the station and other miscellaneous parts to make a station work out in space. 9 inches wide.

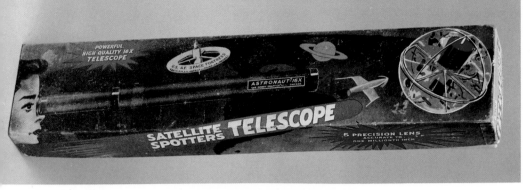

SATELLITE SPOTTER TELE-SCOPE, 1958. A child's plastic telescope to look for satellites.

SPACE SATELLITE MOON RADIO, 1958. A simple crystal radio made in the shape of a satellite. For those unfamiliar with the operation, you grounded one wire, put the earphone in your ear and tuned the crystal until you picked up a radio station. No batteries were necessary.

SPUTNIK 9kt GOLD CHARM, ca. 1958. The charm for a charm bracelet was sold at the Brussels World's Fair in Belgium in 1958.

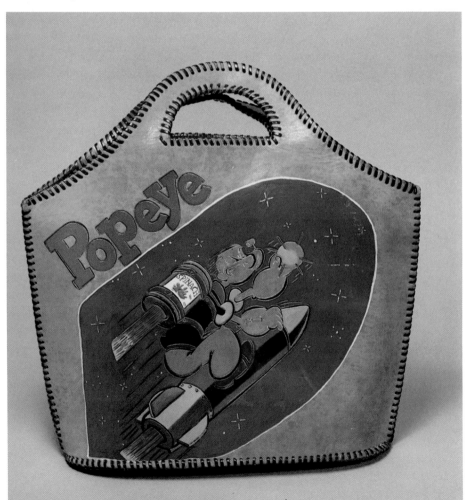

POPEYE SATELLITE HANDBAG, 1958. A most unusual handmade and hand-painted leather handbag with Popeye, powered by spinach, delivering a satellite into space.

FLYING SAUCER SALT & PEPPER SHAKERS, ca. 1959. These salt & pepper shakers were made in the shape of the flying saucers in the movie, "The Mysterians" (MGM-1959).

SPUTNIK BRAND CIGARETTES, ca. 1959. Russia had a great time issuing everyday objects named after the Sputnik. These were sold in the U.S. at the United Nations building's gift shop in New York City.

OFF ON A COMET BY JULES VERNE, March, 1959. A Classics Illustrated comic book.

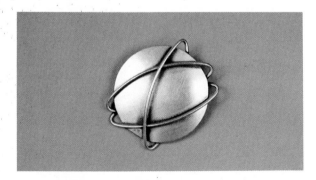

SATELLITE JEWELRY, ca. 1959. A silver pin representing a satellite orbiting the World.

SPACE BOOK, 1960. *First Men To The Moon* by Werner Von Braun (1960). Von Braun tells how we will get to the moon.

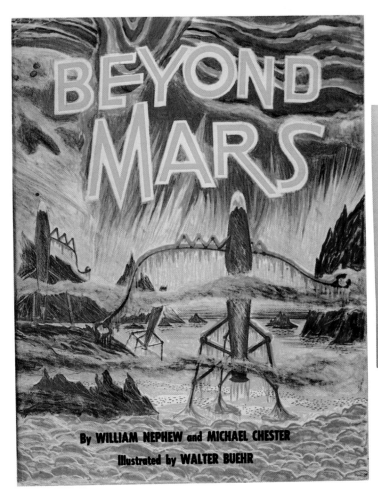

ROCKETS, JETS AND MISSILES, December, 1960. A Classics Illustrated Special Edition comic book with a story about the new group of astronauts.

SPACE BOOK, 1960. *Beyond Mars* by Nephew & Chester (1960). Where will space exploration continue after we explore Mars.

ROCKET JET PLAYGROUND RIDE, ca. 1960. These rides for small children were made of cast aluminum and sat atop a large spring attached to a plate in the ground. Toddlers could make believe that they were flying the latest rockets.

ATOMIC JEWELRY, ca. 1960. Sterling silver cufflinks and tie clip capture the power of the atom. The image of the atom was very popular as it was the symbol of the atomic age—the era of cheap power.

SPACE MISSILE FIRECRACKERS, ca. 1960. A brick (80 packs of 16 firecrackers) label. Now that the U.S. was gearing up to put a man in space, the firecracker makers were pushing space and rocket themes.

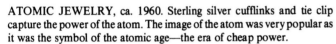

SPACE CHARM BRACELET, ca. 1960. The charms are a telescope, a spaceman, a moon orbiting a globe, a satellite, a rocket, and a shooting star.

SATELLITE CURRENCY, ca. 1960. Each coin has a Sputnik on the face and the caption "The third year of the world in space".

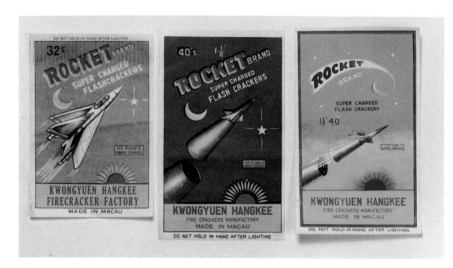

ROCKET BRAND FIRECRACKERS, ca. 1960.

FIRECRACKER LABELS, 1960s. Atomic Missile, Missile and Martian firecracker labels.

SPACE AGE FIRECRACKERS, ca. 1960. A string of 120 firecrackers.

ROCKET RADIO, 1960. A crystal radio made in the shape of a rocket.

SPUTNIK BRAND AFTER SHAVE CREAM, ca. 1960. This may have been an export item since the other side of the box is written in English.

КРЕМ ПОСЛЕ БРИТЬЯ
Спутник

FLYING SAUCER LAMP, ca. 1960. An incredible 3 foot tall flying saucer lamp with lights above, below and around the center. It predates the lunar lander but evokes its image. Most moon landing vehicles of the 1950s and 1960s used the padded foot design.

CHRISTMAS ORNAMENT, ca. 1960. Santa rides a rocket to make his Christmas deliveries.

TOOTHPICK DISPENSER, ca. 1960. These lovely little rockets might have stood on each table in a restaurant in the days when small items did not "walk off" so quickly. A gentle push down dispensed one toothpick

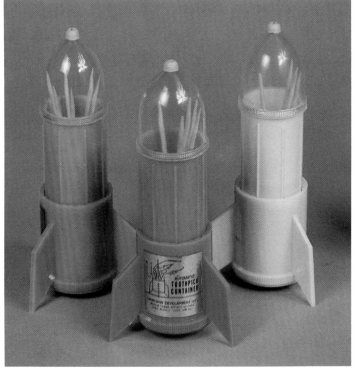

ONAUT HALLOWEEN COSTUME, ca. 1961. Made by Halco.
hard to find costume. Astronauts were "hot" and it was reasonable
ect lots of kids to want to be NASA astronauts for

Halloween. The demand never materialized and the costum
discontinued soon thereafter. Another version was made by Colle
after we landed a man on the moon.

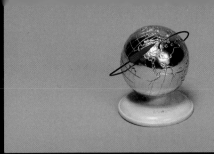

ROCKET DESK ORNAMENT, 1961. A tiny Soviet made model of the rocket that carried Yuri Gagarin, first man in space, soaring above the earth.

VENTURES COMIC BOOK, December, 1961. An action
nture with a realistic touch.

RUSSIAN ENAMEL LAPE
PIN, 1961. The pin was given
those working on or connecte
with the Vostok 2 launch
August 6, 1961.

REDSTONE ROCKET MODEL, ca. 1961. A Revell model kit in
box. Model building was popular and numerous kits were issued w:
space stations, rockets and satellites.

SPACE MEDALLION WITH ROCKET AND SATELLITES,
1961. The medallion is mounted as a paperweight upon a piece of mar
Made in Italy by "C.L. Med".

U.S. ASTRONAUT HALLOWEEN COSTUME, ca. 1961. Made
Ben Cooper, this was Cooper's attempt to capture the interest created

AUTOGRAPHED PHOTO OF ALAN SHEPARD, 1961. The astronauts were just becoming our new heros and they were accessible to the autograph seeking public. Once NASA's publicity people recognized the demand, the astronauts were rephotographed in spacesuits with official props so that the photos would add to the promotion of the space program. As time went on and the demand for the astronaut's autographs increased, the astronauts began to use the Autopen. The Autopen could reproduce their signature many times over. Autopen signatures are worth less than the original signed pieces. These early civilian photos are extremely difficult to find with autographs.

AUTOGRAPHED PHOTO OF GUS GRISSOM, 1961. Gus Grissom later died in the Apollo 1 fire. Grissom's autograph is one of the most actively sought of all of the astronauts.

AUTOGRAPHED PHOTO OF GORDON COOPER, 1961.

AUTOGRAPHED PHOTO OF WALLY SCHIRRA, 1961.

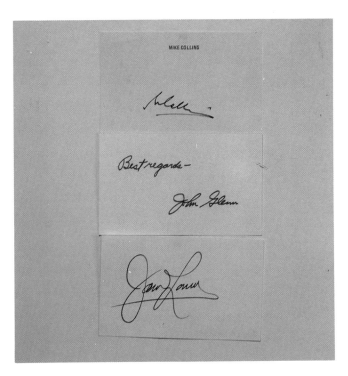

AUTOGRAPHED PHOTO OF DEKE SLAYTON, 1961.

AUTOGRAPHS OF THE ASTRONAUTS, 1960s. Mike Collins, John Glen and James Lovell. Some of the astronauts will not sign autographs, others will only sign items inscribed to a particular person so that the "autograph value" will be reduced if the piece is offered for resale.

POSTCARDS, 1960s. Hundreds of varieties of postcards were issued with space themes. Three variations—a real photo card of Man's Landing on the Moon, a 3-D card of Man Approaching the Moon and a 1961 Russian postcard honoring Yuri Gagarin. The Soviet cards have been difficult to obtain in the past but with the opening of the borders, more and more material appears in the western market each year.

POSTAGE STAMPS, 1960s. Foreign postage stamps were created honoring the different astronauts and space achievements. These are facsimiles. When duplicates are compared, the post marks are identical and actually printed on the stamps.

CUFFLINKS WITH THE TIROS V SATELLITE, 1962. RCA gave this set of cufflinks to executives who helped put this satellite into space.

POSTCARDS, 1960s. Three variations—a real photo card of Apollo 11's lunar lander returning to the command module after leaving the Moon, the fiery liftoff of a Gemini Titan rocket on June 3, 1964 and a 1965 postcard of Werner Von Braun watching a rocket launch. Postcard collecting is very popular and the variety of space cards seems endless and the prices are usually very reasonable. Be warned that there are reproduction cards that were created to be used and do not have collector value.

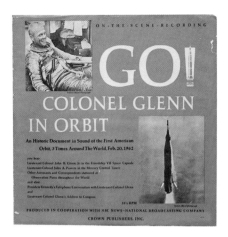

"GO! COLONEL GLENN IN ORBIT" RECORD, 1962. A great longer version of John Glenn's historic flight and the President's telephone conversation with Glenn. 33 rpm.

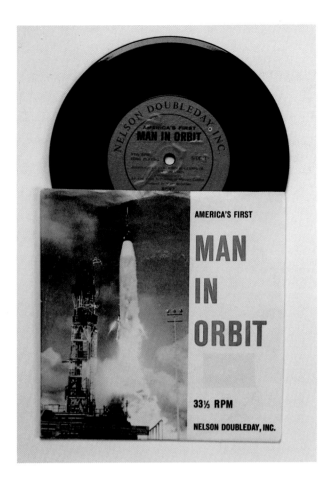

"GLENN REPORTS, 'IT'S BEAUTIFUL'", February 20, 1962. News-papers were filled with loads of information about each flight. They are an incredible source of data about the men and their flights. The only disadvantage is that newsprint is printed on a wood pulp paper that is destined to disintegrate over time. For archival purposes, they should not be exposed to light and a light dusting with baking soda may help to neutralize the acid created by the aging process.

"MAN IN ORBIT" RECORD, 1962. If you missed the launch on television or radio, you could have this recording to hear everything that took place on that historic day. 33 rpm. This was issued by a science object-of-the-month club.

MERCURY CAPSULE CHARMS, ca.1962. Small plastic charms only an inch tall.

FIRST DAY COVERS, 1962. One FDC honors Project Mercury and the other honors the flight of Wally Schirra on October 3, 1962.

AUTOGRAPHED FIRST DAY COVER, February 20, 1962. This was John Glenn's flight. Glenn's autograph is often made with the Autopen. It may take an expert to tell the difference. Earlier pieces are often signed many years after the event.

"GREAT SCOTT", May 25, 1962. The Daily News reports that Scott overshot the mark and was hard to find.

MYSTERY IN SPACE, June, 1962. Science fiction space adventures sold comic books.

SPACE ADVENTURES COMIC BOOK, July, 1962.

SATELLITE BANK PLAY MONEY, ca. 1963. Each coin has an astronaut on its face.

MERCURY CAPSULE LUNCHBOX, ca. 1963. One side shows the capsule and astronaut and the other side shows the rocket lifting off. Made by Thermos. This lunchbox was taken off the market after one year since the cutaway capsule design was taken from a National Geographic article without permission.

FACSIMILE AUTOGRAPHED PHOTO, 1964. The Gemini astronauts. A facsimile autograph is usually printed with the photograph or print. It generally adds no extra collector's value.

AUTOGRAPHED PHOTO OF CHARLES CONRAD, JR., 1965.

MEDALLION FOR THE FIRST GEMINI SPACE WALK, 1965. 2.5 inches in diameter

SPACE CLOCK, ca. 1964. An electric clock with the hour hand as a Nike style missile, the minute hand as a satellite and the second hand as a space capsule.

FAREWELL GEMINI PARTY TICKETS, December 10, 1966. A skating arena local to the launch area threw a party for the astronauts. The last Gemini flight had gone up in November, 1966.

SIX FOOT ROBOT, ca. 1966. This large robot bends and blinks and communicates via an eight track tape deck. It was expected that robots would ride on the most dangerous space missions and the image of a man-like robot evolved. They always caught your attention.

CAPTAIN ASTRO LUNCHBOX, ca. 1966. Great graphics adorn this lunch box, Captain Astro was created in 1960 by the Etch-a-Sketch Company. Made by Ohio Art Metal.

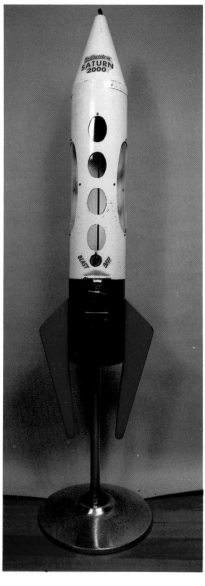

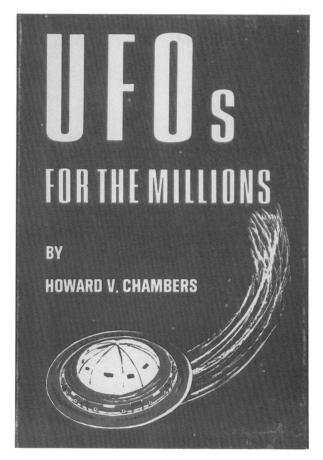

FLYING SAUCER BOOK, 1967. *UFOs For The Millions* by H.V. Chambers (1967). Almost everything that you always wanted to know.

ROBOT GUMBALL MACHINE, ca. 1966. The talking gumball machine made by the Northwestern Co. of Morris, Illinois used some of the parts from other machines. It stood 4 feet tall.

ROCKET GUMBALL MACHINE, ca. 1966. This Saturn 2000 made by the Northwestern Co. of Morris, Illinois stood 5′ 4″ tall.

LOST IN SPACE LUNCHBOX, ca. 1967. The show ran from 1965 to 1968 and its most endearing personality was a robot. There are mountains, craters and a flying saucer on the reverse side. The show was based upon a Robinson Crusoe-type plot and involved a family aptly named the Robinsons. While not officially a "race for space" item, the show helped keep the public's interest in space exploration while we prepared to land a man on the moon.

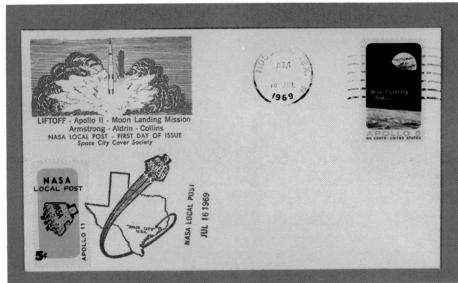

LIFTOFF OF APOLLO 11, July 16, 1969. A first day cover created for the liftoff of the flight that will land a man on the moon.

APOLLO 8 WINE JUG, Italian Barsottini red win in honor of our first flight around the moon, 1968.

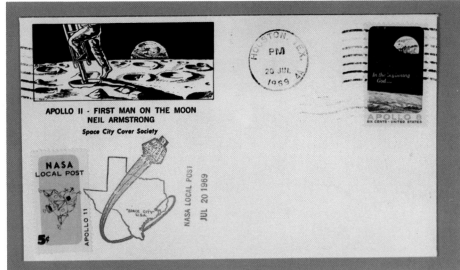

LANDING OF APOLLO 11, July 20, 1969. A first day cover created for the landing of Apollo 11 on the moon.

LANDING OF APOLLO 11, July 20, 1969. A first day cover created for the landing of Apollo 11 postmarked from Tranquility, New Jersey.

"ASTRONAUTS SWING INTO MOON ORBIT", July 20, 1969. The New York Times prepares us for the landing.

PHOTOGRAPH ON THE MOON, July 20, 1969. This is an original print (11″ x 14″) of one of the first photographs taken by Neil Armstrong on the moon with the specially modified Hasselblad camera. It is of Buzz Aldrin assembling scientific apparatus. Copy prints from this were sent to the press and wire services throughout the world.

"MEN WALK ON THE MOON", July 21, 1969. The New York Times gives front page coverage to "All the news that's fit to print".

PHOTOGRAPH ON THE MOON, July 20, 1969. This is a copy print
(8″ x 10″) of one of the most famous color photographs taken by Neil
Armstrong of Buzz Aldrin with the reflection of Armstrong and the lunar
lander reflected in the visor. The photo is autographed by Buzz Aldrin.

LANDING OF APOLLO 11, July 21, 1969. A first day cover created in honor of the landing of Apollo 11 and postmarked Jordell Bank, Cheshire, England, one of the listening stations for the Apollo flights.

RETURN OF APOLLO 11, July 24, 1969. A first day cover created for the return of Apollo 11 and postmarked Western Somoa, one of the listening stations for the Apollo flights.

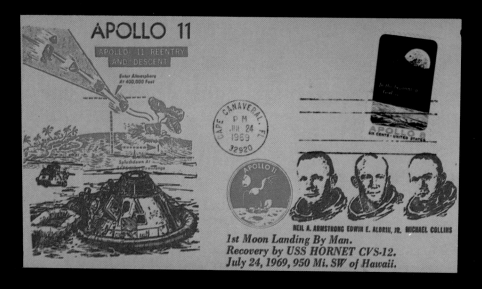

RETURN OF APOLLO 11, July 24, 1969. A first day cover created for the return of Apollo 11.

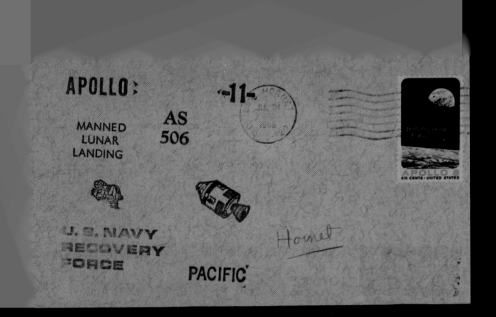

RECOVERY OF APOLLO 11, July 24, 1969. A first day cover created for the recovery of Apollo 11, postmarked on the U.S.S. Hornet, the ship that picked up the astronauts.

FIRST DAY COVER FOR APOLLO 11, 1969. A first day cove created for Apollo 11 from Granada.

COMMEMORATIVE STAMP FOR APOLLO 11, 1969. Special stamp created for the return of Apollo 11 from Magyar. Numerous countries issued commemorative stamps honoring the Apollo 11 flight.

COMMEMORATIVE STAMP FOR APOLLO 11, 1969. Specia stamp created for the return of Apollo 11 from Liberia.

COMMEMORATIVE STAMP FOR APOLLO 11, 1969. Special stamp created for the return of Apollo 11 from Romania.

COMMEMORATIVE STAMP FOR APOLLO 11, 1969. Special stamp created for the return of Apollo 11 from the Republic Togolaise.

FIRST MEN TO THE MOON PAINTING, 1969. An oil painting by a talented unknown artist. Many artists recorded their impression of the first landing on the moon. This is an exciting field of collecting as the art work is space age folk art. 26″ X 32″.

MOON LANDING PAINTING, 1969. An oil painting on masonite by an unknown artist of the Lunar lander and command module approaching the moon with earth in the background. 16″ x 24″.

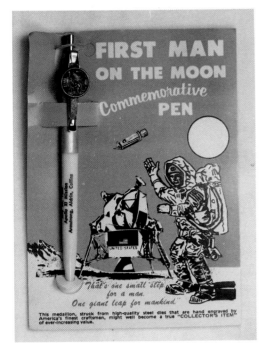

OFF TO THE MOON SCULPTURE, 1969. A copper sculpture by Michael B. Madeclt of the moon rocket blasting off into space on its journey to the moon.

FIRST MAN ON THE MOON COM-MEMORATIVE PEN, 1969. Inexpensive pen with a small medal attached that, according to the attached card "...might well become a true 'COLLECTOR'S ITEM' of ever-increasing value". It is hard to gauge the actual collector's interest in these "made for collector" items. The commemorative items destined to be the most collectible are those that were probably too tacky to attract much collector interest. The most available items are the inexpensive metal coins or medallions made to commemorate each and every flight.

FIRST DAY COVER FOR APOLLO 11 STAMP, September 9, 1969. A first day cover created for the stamp commemorating the landing of Apollo 11.

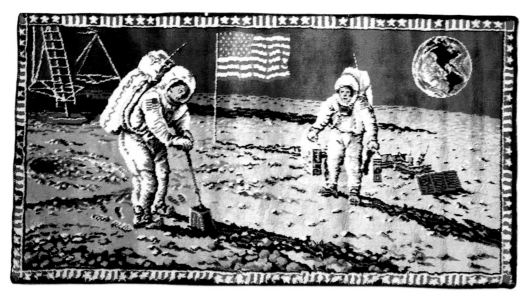

FIRST MAN ON THE MOON
COMMEMORATIVE RUGS,
1969. These rugs were made in Italy
and sold in the U.S. They are 37″ x
20″ and could be called very "tacky".

FIRST MAN ON THE MOON PLAQUE, 1969. A plaster plaque 15″ in diameter made to hang on a wall.

FIRST MAN ON THE MOON COMMEMORATIVE MEDALLION, 1969. 1″ wide.

FIRST MAN ON THE MOON COMMEMORATIVE MEDALLION, 1969. 2.5″ in diameter.

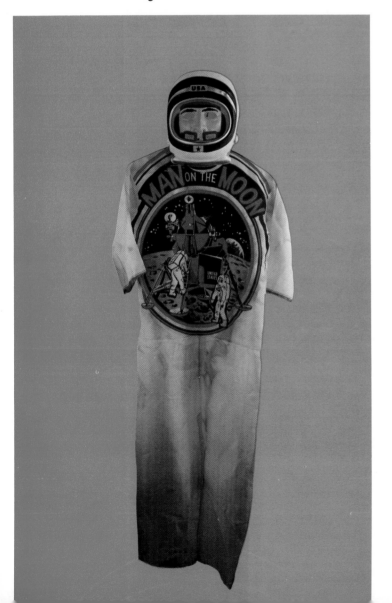

FIRST MAN ON THE MOON COSTUME, 1969. A Halloween costume by Collegeville. This seems like a natural concept for a 1969 Halloween costume but few have ever seen this one. The scorch marks along the front and sides are typical. The silky material coated with a fire resistant coating either reacted badly to a hot iron or prolonged storage.

COMMEMORATIVE MEDAL FOR APOLLO 11, 1969. This nice high relief medallion was created in honor of the Apollo 11 flight. Commemorative medals were made by the thousands but some, like this one, exhibit a degree of craftsmanship above the ordinary.

APOLLO COMMEMORATIVE PLATE, 1969. There are numerous variations of first man on the moon commemorative plates. This version is plastic and sports a real photo of our first man on the moon. It was not made in a limited edition and there are probably less of these around than the limited edition varieties.

MOON GLOBE, 1969. A small globe of the moon showing the craters and landing site of Apollo 11.

Opposite page:

MOON LANDING BEDSPREAD, 1969. A great bedcover with an image of the Lunar lander and first man on the moon. Twin bed size.

FIRST MAN ON THE MOON RECORD & BOOK, 1969. Time-Life
came out with this handsome book and record in a slipcase and entitled it,
"To The Moon". 33 rpm.

FIRST MAN ON THE MOON COMMEMORATIVE RECORD, 1969. 45 rpm.

MAN ON THE MOON RECORD, 1969. Another version of the tapes and commentary of our first landing on the moon. 33 rpm.

FLIGHT TO THE MOON RECORD, 1969. Another version of the tapes and commentary of our first landing on the moon. 33 rpm.

SOYUZ-5 FIREWORKS, ca. 1969. A label from a box of skyrockets.

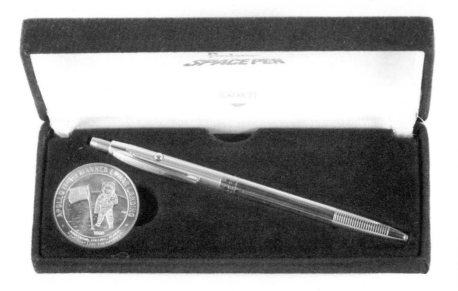

FISHER SPACE PEN & REFILL, 1969. The Fisher pen company made a ballpoint pen that was used in space. It used a pressurized refill and did not rely upon gravity to work. Their packaging hitched on to the popularity of the space program.

SPACE CAPSULE CARNIVAL RIDE, ca. 1969. A wonderful space capsule from a carnival ride. The savvy astronaut holds a "U.S. Astronaut" decorated styrofoam space helmet made about 1965.

PARKER PEN, 1970. The T-1 integral nib fountain pen was made of titanium and was created in honor of our landing a man on the moon. Parker only made them for one year and due to the high cost of working with titanium, lost money on every pen produced and sold.

FIRST ON THE MOON

A VOYAGE WITH
NEIL ARMSTRONG
MICHAEL COLLINS
EDWIN E. ALDRIN, JR.

EPILOGUE BY ARTHUR C. CLARKE

SPACE BOOK, 1970. *First On The Moon* by Armstrong, Collins & Aldrin (1970). Here are the three astronauts' own stories.

MOON BOOTS, ca. 1970. Those big bulky apres-ski boots became Moon Boots.

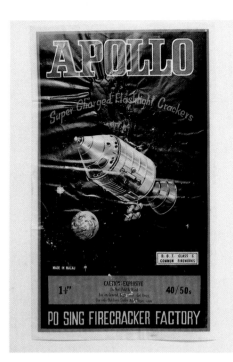

APOLLO FIRECRACKERS, ca. 1970. A brick (40 packs of 50 firecrackers) label.

MATCH PACK, 1971. From a dance near the Cape honoring the astronauts who would fly aboard Apollo 15.

APOLLO COMMEMORATIVE GLASS SET, ca. 1971. A set to sip lemonade on those hot Florida afternoons honoring Apollos 11 through 14. It is unusual to cover only 4 flights, but perhaps there were other sets.

OIL PAINTING, ca. 1971. "That's How It Felt To Walk On The Moon"
by Alan Bean, From the limited edition print copyright © The Greenwich
Workshop, Inc., Trumbull, CT. Alan Bean paints from actual experience
on the moon.

OIL PAINTING, ca. 1971. "Helping Hands" by Alan Bean, From the limited edition print, copyright © The Greenwich Workshop, Inc., Trumbull, CT. Alan Bean is one of the 12 men to have walked on the moon. He was aboard Apollo 12. After retiring from NASA, he took up painting as a full time avocation. This shows Alan Bean and Pete Conrad walking together on the moon. Bean's paintings are realistic and evocative.

CERNAN SALUTES FLAG AT APOLLO 17 LANDING SITE

AUTOGRAPHED PHOTO OF GENE CERNAN, 1972.

APOLLO-SOYUZ CARTOON, July 15, 1975. An original pen and ink drawing by Stanley Franklin showing the U.S. capsule serving short order food to the Russians. A notation indicates that it appeared in "The Sun".

APOLLO COMMEMORATIVE MATCH PACK, ca. 1975. RCA produced this interesting souvenir of the Apollo program.

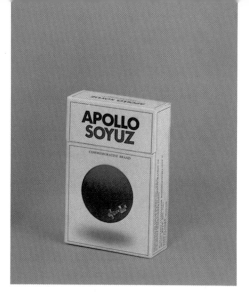

APOLLO-SOYUZ CIGARETTES. 1975. Cigarettes developed by Philip Morris, Inc. and made in Moscow at the Yava factory "in commemoration of U.S./Soviet space cooperation".

MINUTEMAN LAUNCH, 1979. A photograph of dual minuteman missiles being launched. Signed by the commander of the 1st Strategic Aerospace Division. These types of missiles were our defense against the Soviet threat.

10TH ANNIVERSARY MEDALLION FOR APOLLO 11, 1979. 1.75″ in diameter.

SPACE SHUTTLE ELECTRONIC COMPONENT, ca. 1980. Miscellaneous parts were ordered and never used. This one comes with a certificate of authenticity and estimate of value.

10TH ANNIVERSARY FIRST DAY COVER FOR APOLLO 11, 1979.

PRESS PASSES FOR SPACE SHUTTLE LAUNCHES, 1981. These were for the first two launches.

SPACE SHUTTLE DIGITAL CLOCK, ca. 1981.

GREETING CARDS, 1981. These were issued by Hallmark right after the first Space Shuttle flight. Copyright © Hallmark Cards.

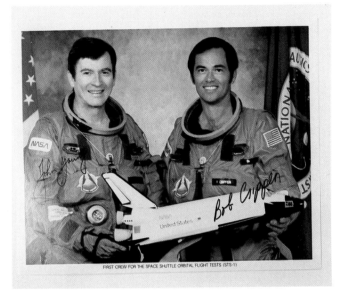

AUTOGRAPHED PHOTO OF THE CREW OF THE FIRST SPACE SHUTTLE FLIGHT -STS-1, April,1981. Pilots were Crippen and Young.

MAGICIAN'S PROP, ca. 1981. A popular magician's trick known as "Run Rabbit Run" was updated to "Space Trek". The object was to find the Martian as he moved from rocket to rocket. The magician doesn't notice the Martian moving but the young audience sees the Martian move from side to side and really gets involved trying to point it out to the Magician.

PRESS KIT, 1981. Press kits were issued by Rockwell International after the successful completion of the first space shuttle flight. They were public relations items and schools ordered them by the thousands. They introduced many children to the space program.

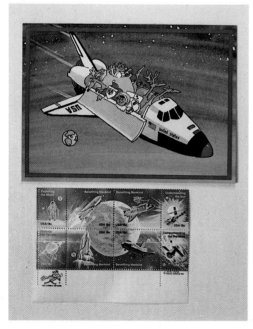

GREETING CARD AND POSTAGE STAMPS, 1981. The Christmas card was issued by Smithsonian right after the first Space Shuttle flight. The postage stamps were issued before the first shuttle flight extolling the benefits to be received from the flights.

AUTOGRAPHED PHOTO OF CREW OF THE SECOND SPACE SHUTTLE FLIGHT -STS-2, November, 1981. Pilots were Truly and Engle.

NEWSWEEK MAGAZINE, April 27, 1981.

SHUTTLE POSTER, 1981. The Postal Service issued the shuttle stamp in honor of the first Space Shuttle flight.

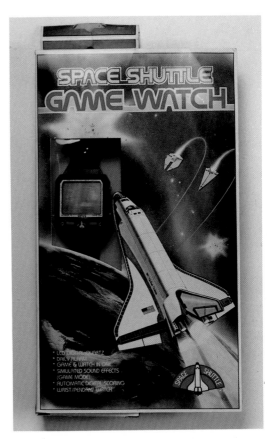

SPACE SHUTTLE TILE PIECE, ca. 1981. An enterprising fellow found a dumpster full of discarded space tiles used to cover the outside of the Space Shuttle. He cut them into tiny pieces and produced this interesting souvenir of the first shuttle.

SPACE SHUTTLE GAME WATCH, ca. 1981. Now one could tell time and play a space shuttle game. In typical game fashion, you had to get the shuttle from point A to point B without being blown up.

AUTOGRAPHED PHOTO OF THE CREW OF THE THIRD SPACE SHUTTLE FLIGHT-STS-3, March, 1982. Pilots were Fullerton and Lousma.

SPACE SHUTTLE WINE 1982. *Space Shuttle Red* and *Space Shuttle White* wines made by Bully Hill Vineyards of New York State. Bully Hill's owner is a great promoter of his New York State wines and wanted a special edition in honor of the shuttle flights.

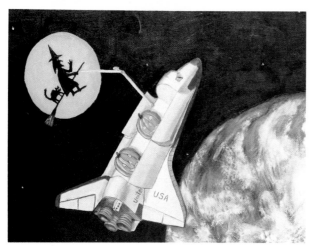

SPACE SHUTTLE PAINTING, 1984. An oil painting by S. Schneider of the Shuttle putting the witch in front of the moon for Halloween. 14.5" x 18".

SPACE SHUTTLE PAINTING, 1984. An oil painting by S. Schneider of the Shuttle being launched. Note the shuttle tile-like frame and the dark main fuel tank. NASA found that it could cut thousands of pounds of weight off the shuttle by not painting the main fuel tank with a second coat of paint. 14.5" x 18".

AUTOGRAPHED PHOTO OF THE CREW OF THE SPACE
SHUTTLE, February, 1984. Ron McNair was later killed aboard the
Challenger in January, 1986.

SPACE SHUTTLE PAINTING, 1985. An oil painting by S. Schneider of the Shuttle out in space and Halley's comet approaching earth. 11″ x 14″.

AUTOGRAPHED PHOTO OF THE CREW OF THE SPACE SHUTTLE 51-D, 1989.

RUSSIAN SPACE SHUTTLE WATCH, ca. 1989. Russia built and tested its own space shuttle. This watch was to commemorate the anticipated flight. With the breakup of the Soviet Union, it is believed that it will never fly commercially.

SPACE SHUTTLE PAINTING, 1990. An acrylic painting by Neil DePinto of the Shuttle Atlantis during liftoff. DePinto's style is photo realism. One can almost feel the heat of the rocket's exhaust. Size is 15″ x 20″.

THE HUBBLE SPACE TELESCOPE. April 25, 1990. Photo of the
Hubble Space Telescope being released from the Shuttle Discovery—
STS-31.

THE HUBBLE SPACE TELESCOPE. 1990. Illustration of the Hubble
Space Telescope over earth by artist Dana Barry. Space art enhances any
space item collection.

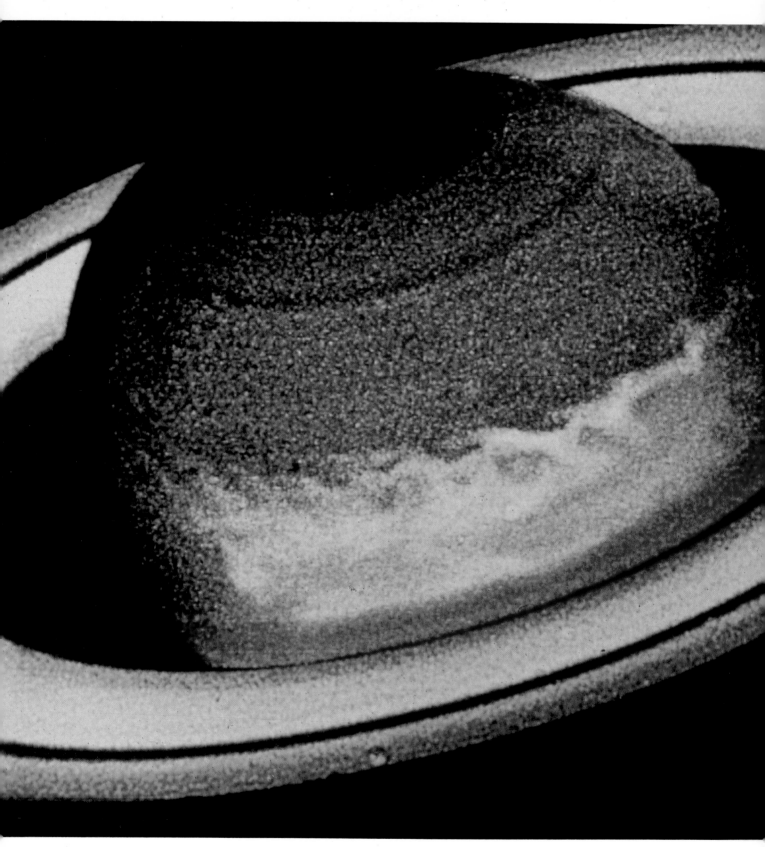

PHOTOGRAPH OF SATURN, 1990. This photo was taken through the Hubble Space Telescope. For the collector who collects photographs, few capture the feel of being there like those shot through the Hubble Space Telescope.

PHOTOGRAPH OF JUPITER, 1991. This photo was taken through the
Hubble Space Telescope.

Bibliography

Barbour, J. *Footprints on the Moon*. Associated Press, NY. 1969
Bryan, C. *The National Air & Space Museum*. Abrams, NY. 1979
Curran, D. *In Advance of the Landing*. Abbeville, NY. 1985
Hake, T. *Space Collectibles Auction Catalog*. Hake, PA. 1984
Kitahara, T. *Wonderland of Toys-Tin Toy Robots*. Shinko Music Pub, Tokyo. 1983.
Malone, R. *The Robot Book*. Push Pin, NY. 1978
Miller & Hartmann. *In the Stream of Stars*. Workman, NY. 1990
Millspaugh, B. *Aviation & Space Science Projects*. Tab, PA. 1992

Sansweet, Stephen. *Science Fiction Toys and Models*. Starlog Press, New York. 1980.
Schneider, Stuart *Halley's Comet-Memories of 1910*. Abbeville, NY. 1985
Singer, Leslie. *Zap*. Chronicle Books, San Francisco. 1991.
Swanson, Glen *Quest Magazine*. CSpace, MI. 1992
Time-Life. *Life in Space*. Time-Life, Chicago. 1983
Tumbusch, T.N. *Space Adventure Collectibles*. Wallace-Homestead, PA. 1990.

Resources

Astrophile, Space Topic Study Unit of the American Philatelic Society, c/o Bernice Scholl, P.O. Box 2579, Marathon Shores, FL 33052. Great source of information for collectors of Space Stamps and first day covers.

Countdown Magazine, P.O. Box 216, Athens, OH 45701. Magazine for those interested in the current space program.

Final Frontiers Magazine, 1516 W. Lake St., Ste.102, Minneapolis, MN 55408

NASA, LBJ Space Center, Houston, TX, 77058. Source of information and current astronaut autographed photos.

NASA Jet Propulsion Laboratory, 4800 Oak Grove Dr., Pasadena, CA 91103

NASA Space Link, Computer bulletin board. Phone number (205) 895-0028 or access through your local "Internet" system (address is "spacelink.msfc.nasa.gov 128.158.13.250"). Located in Huntsville, AL. At Welcome screen, hit Return and type "NEWUSER" as your name and as your password. Enter # of lines your computer can display (usually 24) and you are in the system. System is specifically designed for teachers and has a vast NASA database.

Quest Magazine, The History of Spaceflight Magazine, P.O. Box 9331, Grand Rapids, MI 49509. Enthusiastic editor who loves the subject matter.

Robot World & Price Guide, c/o Ernie Mannix, P.O. Box 184, Lennox Hill Station, New York, NY 10021. Great for the robot and space toy enthusiast.

Society for the Advancement of Space Activities, P.O. Box 192, Kents Hill, ME 04349. Interest in real space activities.

Space Patch Club, P.O. Box 17310, Pittsburgh, PA, 15235

Space Telescope Science Institute, 3700 San Martin Dr., Baltimore, MD, 21218. The institute conducts scientific exploration with the Hubble Space Telescope.

The Greenwich Workshop, Inc., P.O. Box 393, Trumbull, Ct. 06611 (source for Alan Bean lithographs)

National Space Organizations

American Institute for Aeronautics and Astronautics, 370 L'Enfant Promenade SW, Washington, DC 20024

American Space Foundation, 111 Massachusetts Ave. NW, #200, Washington, DC 20001

The Campaign for Space, P.O. Box 1526, Bainbridge, GA 31317

The L-5 Society, 1060 Elm St., Tucson, AZ 85719

National Space Society, 600 Maryland Ave. SW, #203, Washington, DC 20024

SPACECAUSE, 922 Pennsylvania Ave. SW, Washington, DC 20003

SPACEPAC, 2801 B. Ocean Blvd., #S, Santa Monica, CA 90405

The Planetary Society, P.O. Box 40185, Santa Barbara, CA 93140

U.S. Space Foundation, P.O. Box 1838, Colorado Springs, CO 80901

INDEX

PRICE GUIDE
to
Collecting the SPACE RACE

©1993

Using This Guide

The values listed represent, in the opinion of the author, what an item **which is in extra fine working condition** might be valued at by the knowledgeable collector. The price paid for any particular item will naturally vary by geographical location, and will be also be affected by the eagerness of the buyer, willingness of the seller, whether purchased at a retail shop, antique show, or flea market.

Items with broken, missing, or non-stock replacement parts are worth less than the values listed. Rare or desirable items in excellent or mint condition or toys with their box can be worth several times the values indicated. For more information about valuing items, refer to the chapter on Valuing Space Memorabilia.

Ultimately the price paid for any particular item depends on the buyer and seller. The author does not claim to be the final authority on prices and assumes no responsibility for financial loss or gain based on the use of this guide.

Position Code

T = Top	TRC = Top right center
C = Center	BL = Bottom Left
B = Bottom	BC = Bottom Center
R = Right	BR = Bottom Right
L = Left	CL = Center left
TL = Top Left	CR = Center right
TC = Top Center	T-B = Top to bottom
TR = Top Right	R-L = Right to Left
TLC = Top left center	

Pg.	Position	Value	Pg.	Position	Value
2		50-100	15		25-40
3		100-200	16		75-120
4		200-300	18		2000-2500
5		100-200	19		5-10
6		300-350	23	T	1000-1200
7		10-20	23	B	150-200
9		650-750	24	TL	75-150
10		20-30	24	BL	45-65
11		75-150	24	TR	45-65
13		300-400	24	BR	65-85
14		75-100	25	TL	75-125